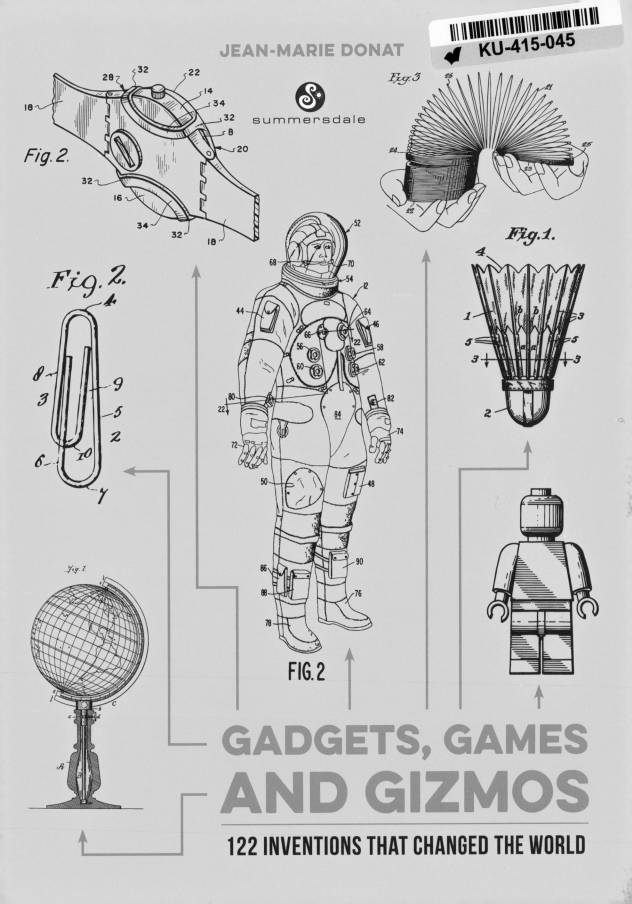

JEAN-MARIE DONAT

summersdale

GADGETS, GAMES AND GIZMOS

122 INVENTIONS THAT CHANGED THE WORLD

GADGETS, GAMES AND GIZMOS

This edition published in 2017 by Summersdale Publishers Ltd

First published by Editions Prisma in 2015

All illustrations (except p.99) courtesy of United States Patent and Trademark Office
http://www.uspto.gov
Illustration on p.99 courtesy of Espacenet

Summersdale Publishers Ltd
46 West Street
Chichester
West Sussex
PO19 1RP
UK

www.summersdale.com

Printed and bound in China

ISBN: 978-1-78685-071-3

Substantial discounts on bulk quantities of Summersdale books are available to corporations, professional associations and other organisations. For details contact general enquiries: telephone: +44 (0) 1243 771107, fax: +44 (0) 1243 786300 or email: enquiries@summersdale.com.

CONTENTS

INTRODUCTION

Humans have always created incredible things, but the recording and controlling of these innovations first emerged in ancient times. The Greek scholar Athenaeus of Naucratis dated the beginnings of intellectual property to the sixth century BC. At that time in Sybaris, a Greek colony in the south of Italy, the creators of culinary recipes were permitted a year's exclusivity for their innovations before handing them over to the community. This practice was established to promote creativity among the city's chefs.

The first patent in a form we would recognise today was registered in Florence in 1421. Attributed to the architect Filippo Brunelleschi, it gave the holder the exclusive use of a hoisting machine for a river barge. Later, in 1474, the Republic of Venice established the first legal framework for the filing of patents.

In 1623, Parliament passed the Statute of Monopolies, the first statutory expression of English patent law. The USA passed its first Patent Act in 1790, and gradually modern systems of patent registration evolved. Today, more than 200,000 patents are registered around the world every year.

Most of these patents concern inventions that are an integral part of our environment – which they shape or even transform. There are so many that we have had to be ruthless in the selection, choosing to highlight those everyday objects which have particularly made their mark on the modern world. From the yo-yo to the space capsule, the parachute to the iPod, the turntable to the computer keyboard, what they all have in common is the fact they were once imagined, then designed, detailed, referenced and finally patented by one of the national bodies for intellectual property.

Illustrated by more than 120 original sketches from the United States Patent and Trademark Office (USPTO) and the European Patent Office (EPO), these designs don't always represent the original invention but sometimes a variation, improvement or accessory; this helps to shed light on the object's history in new ways.

We hope this book will inspire your inner inventor and, with its wealth of amusing and surprising facts, change the way you look at the gadgets, games and gizmos around you.

GAMES AND TOYS

Take a trip down memory lane in this opening chapter, an opportunity to discover the first models of those objects with which we spent the happiest days of our childhood. From Lego bricks to Mickey Mouse, from the Frisbee to the Game Boy, here's a selection of patents that revolutionised forever the world of games and toys, captivating youngsters since 1866. Thanks to these images and anecdotes, you'll long to be a kid all over again!

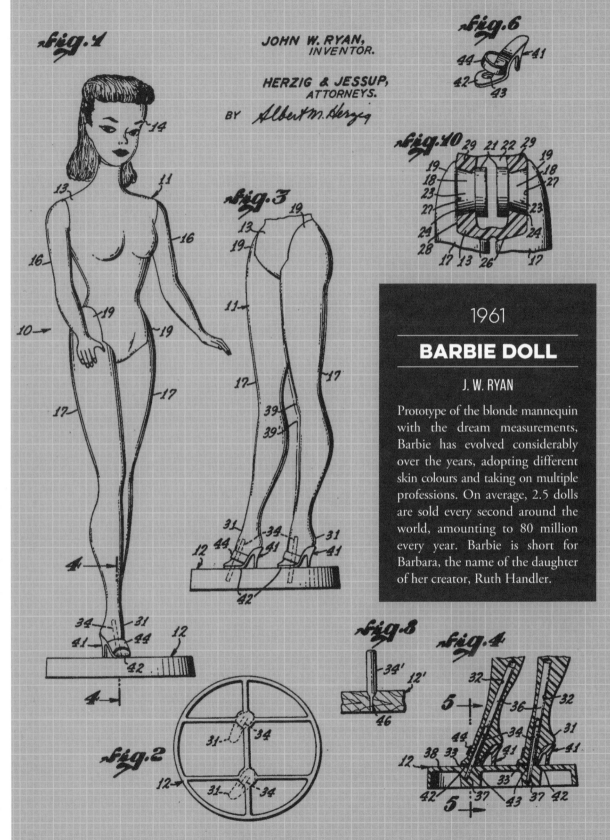

JOHN W. RYAN,
INVENTOR.

HERZIG & JESSUP,
ATTORNEYS.

BY Albert M. Herzig

fig.1

fig.3

fig.6

fig.10

fig.2

fig.8

fig.4

1961

BARBIE DOLL

J. W. RYAN

Prototype of the blonde mannequin with the dream measurements, Barbie has evolved considerably over the years, adopting different skin colours and taking on multiple professions. On average, 2.5 dolls are sold every second around the world, amounting to 80 million every year. Barbie is short for Barbara, the name of the daughter of her creator, Ruth Handler.

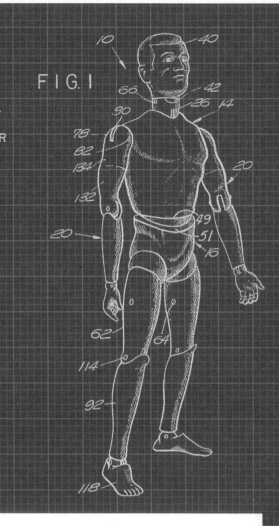

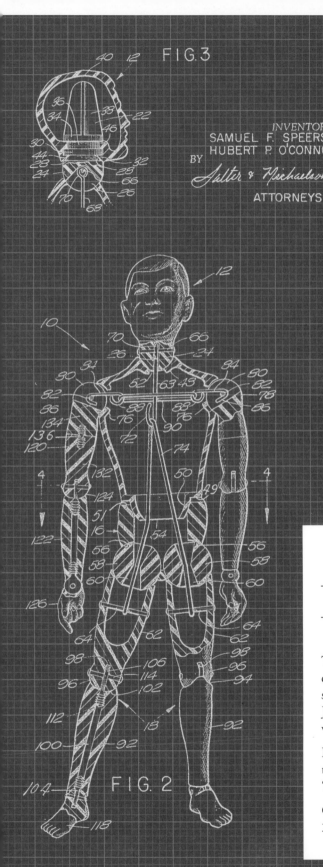

FIG.3

FIG.1

INVENTORS
SAMUEL F. SPEERS
HUBERT P. O'CONNOR
BY
Salter & Michaelson
ATTORNEYS

FIG. 2

1966

G. I. JOE

S. F. SPEERS

The prototypes from Hasbro originally had different names: Rocky (the soldier), Skip (the sailor) and Ace (the pilot). The general name G. I. Joe was finally chosen for this elite team, charged with special missions for the White House…

In the UK, a company called Palitoy created the Action Man as a licenced copy of Hasbro's 'movable fighting man'.

One of the first versions, from 1964, was sold in 2003 for more than $200,000.

Fig. 1

Fig. 2

Fig. 3

1930

MICKEY MOUSE

W. DISNEY

Originally, the Disney mouse was supposed to be named Mortimer. But Walt Disney's wife Lilian thought that sounded too pompous and suggested the name Mickey.

The white gloves, one of the distinguishing features of Mickey's outfit, were adopted for technical reasons: in black-and-white films, the audience could more easily make out the character's hands when they passed in front of a black body.

INVENTOR.

Walter E. Disney

BY *J. S. Bradbury*

186,119

PUPPET DOLL OR SIMILAR ARTICLE

**James M. Henson, Hyattsville, Md., and Jane A. Nebel,
Washington, D.C., assignors to John H. Wilkins Company, Washington, D.C., a corporation of Delaware**

Application October 17, 1958, Serial No. 53,012

Term of patent 14 years

(Cl. D34—4)

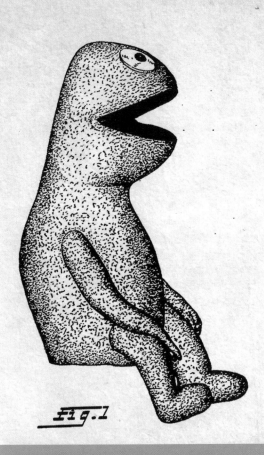

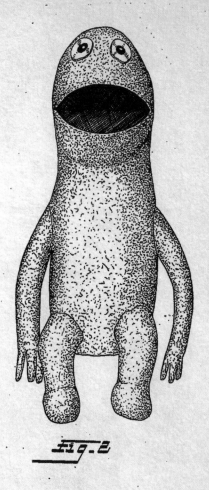

1959

KERMIT THE FROG

M. FLEISCHER

Kermit made his first appearance on 9 May 1955 at
a five-minute puppet show called Sam and Friends.
This pivotal character of The Muppets, which
became a multi-billion-dollar franchise, had a less
than auspicious start – the prototype was made
from a woman's green coat found in a rubbish bin,
with two ping-pong balls for the eyes.

INVENTOR.

Max Fleischer

BY

Kiddle, Margeson & Hornidge
ATTORNEYS.

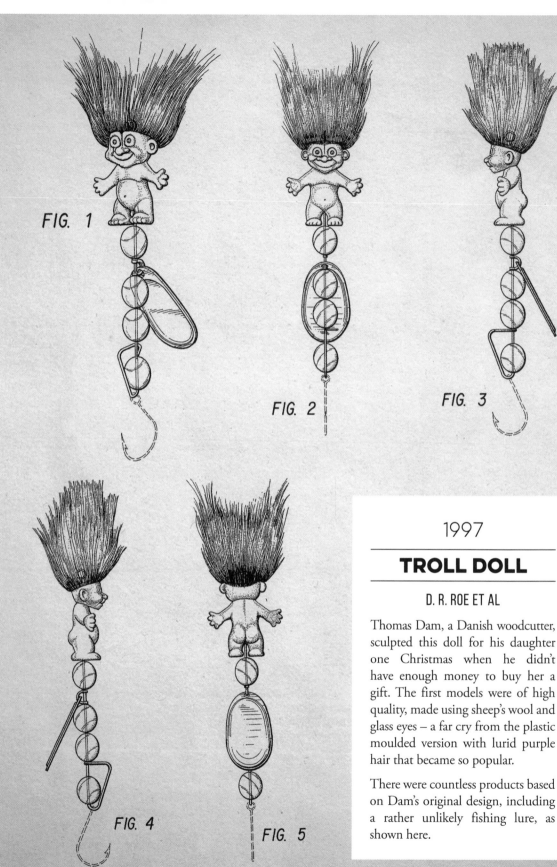

FIG. 1

FIG. 2

FIG. 3

FIG. 4

FIG. 5

1997

TROLL DOLL

D. R. ROE ET AL

Thomas Dam, a Danish woodcutter, sculpted this doll for his daughter one Christmas when he didn't have enough money to buy her a gift. The first models were of high quality, made using sheep's wool and glass eyes – a far cry from the plastic moulded version with lurid purple hair that became so popular.

There were countless products based on Dam's original design, including a rather unlikely fishing lure, as shown here.

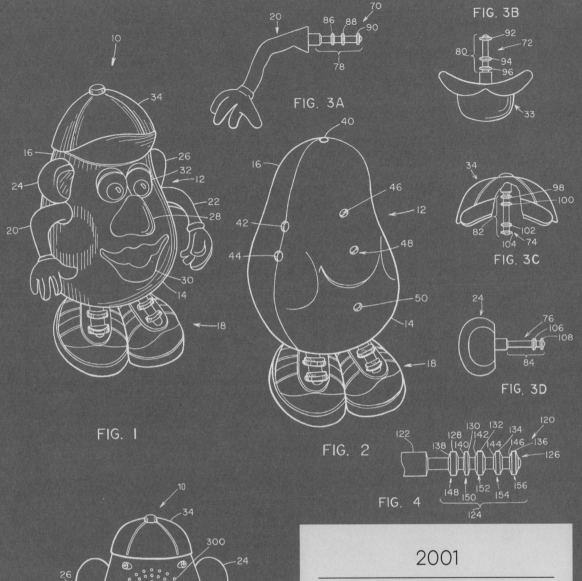

FIG. 1

FIG. 2

FIG. 3A

FIG. 3B

FIG. 3C

FIG. 3D

FIG. 4

FIG. 7

2001

MR POTATO HEAD

DANA A. SILVA

In the beginning, Mr Potato Head was nothing but a set of accessories. Children had to fix the different plastic attributes – hands, feet, nose, eyes, ears, mouth, 'hair' made from pieces of felt, spectacles and a pipe – into real potatoes. The kit didn't come complete with a body until 1964.

In 1987, Mr Potato Head stopped smoking a pipe and became the ambassador of the American Cancer Society.

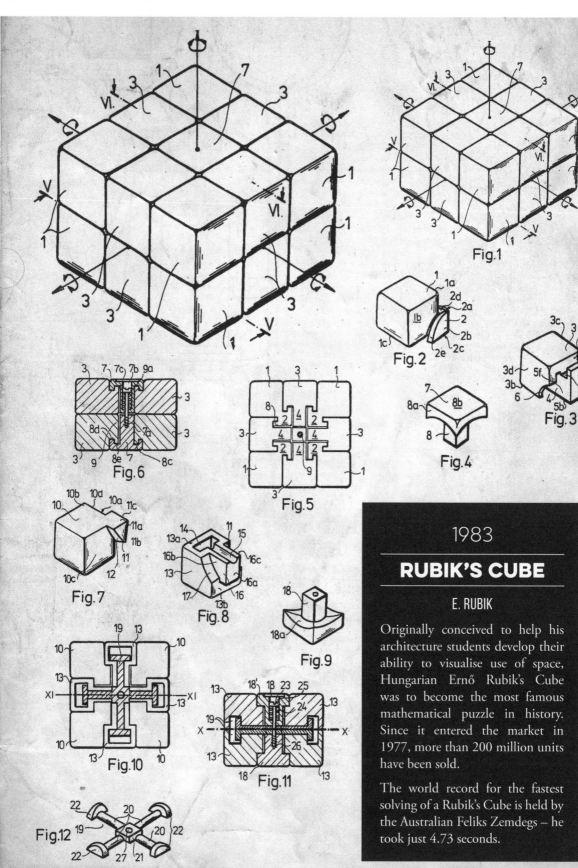

Fig.1

Fig.2

Fig.3

Fig.4

Fig.5

Fig.6

Fig.7

Fig.8

Fig.9

Fig.10

Fig.11

Fig.12

1983

RUBIK'S CUBE

E. RUBIK

Originally conceived to help his architecture students develop their ability to visualise use of space, Hungarian Ernő Rubik's Cube was to become the most famous mathematical puzzle in history. Since it entered the market in 1977, more than 200 million units have been sold.

The world record for the fastest solving of a Rubik's Cube is held by the Australian Feliks Zemdegs – he took just 4.73 seconds.

Fig.1

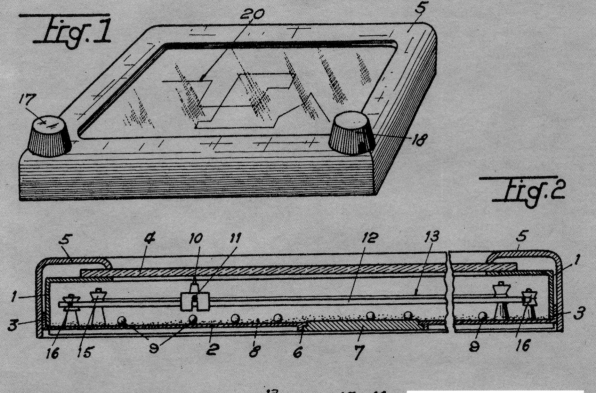

20

5

17

18

Fig.2

5   4   10   11   12   13   5

1   1

3   3

16   15   9   2   8   6   7   9   16

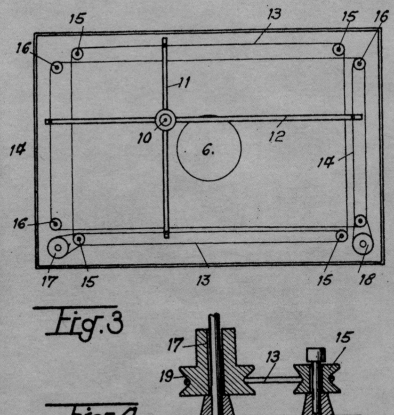

Fig.3

Fig.4

1962

ETCH-A-SKETCH

A. GRANDJEAN

This mechanical drawing board, originally conceived by Frenchman André Cassagnes (though Arthur Grandjean was mistakenly credited on the paperwork), was bought by an Ohio company in 1960. Twisting the knobs on either side moves the stylus, displacing aluminium powder inside, and the screen is wiped clear by shaking it. It was still manufactured in Ohio until 2001, when production moved to China.

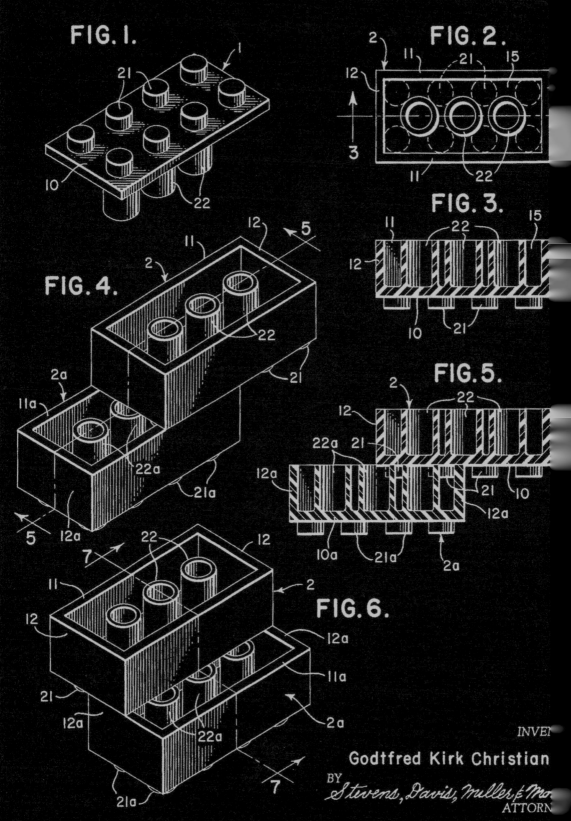

FIG. 1.

FIG. 2.

FIG. 3.

FIG. 4.

FIG. 5.

FIG. 6.

INVEN
Godtfred Kirk Christian
BY
Stevens, Davis, Miller & Mo
ATTORN

14

FIG. 7.

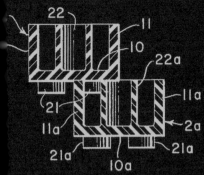

FIG. 8.

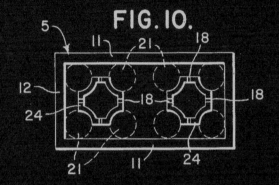

FIG. 9.

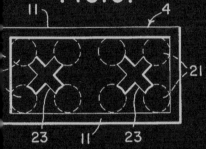

FIG. 10.

FIG. 11

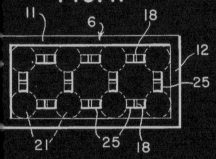

FIG. 12.

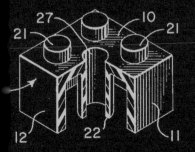

1961

LEGO BRICK

G. K. CHRISTIANSEN

Created by a carpenter with a passion for wooden toys, the Lego brand took its name from the Danish *leg godt*, meaning 'play well'. Some 45 billion building bricks (now plastic) are sold every year. You'd need fewer than that, just 40 billion bricks, stacked one on top of the other, to reach the moon.

A mathematics professor first calculated it was possible to create over 915 million different combinations of the way six 4×2 Lego bricks fit together.

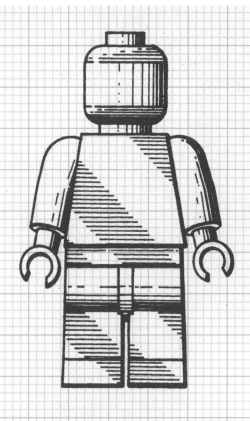

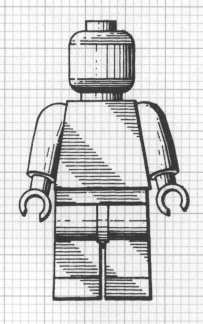

FIG. 1

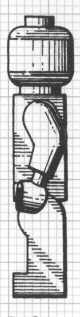

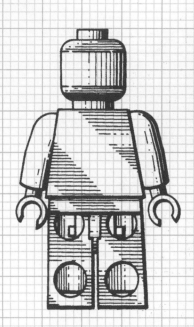

FIG. 2

FIG. 3

1979

LEGO MINIFIGURES

G. K. CHRISTIANSEN

The minifigures made their first appearance in the Lego universe in 1975. The height of four Lego bricks, they had to wait three years before their arms and legs were fully articulated.

According to the brand, there are more than 3,600 models, and approximately 3.7 billion units have been produced since 1978.

FIG. 4

FIG. 5

FIG. 6

FIG. 7

FIG. 8

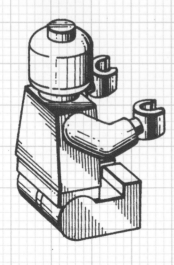

FIG. 9

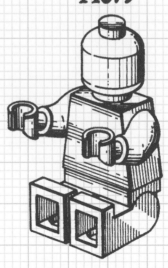

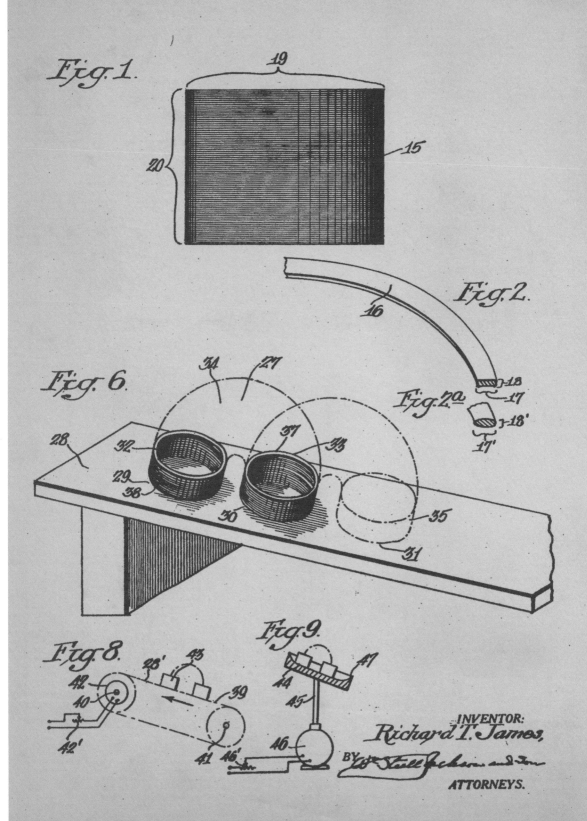

Fig.1.

19

15

20

16

Fig.2.

Fig.6.

34 27

Fig.2a.

18
17
18'
17'

28

32
29
38

31
33

50

35

31

Fig.8.

Fig.9.

28'
42
40

43
39

41
44
45

42'
41 46'
46

INVENTOR:
Richard T. James,
BY
ATTORNEYS.

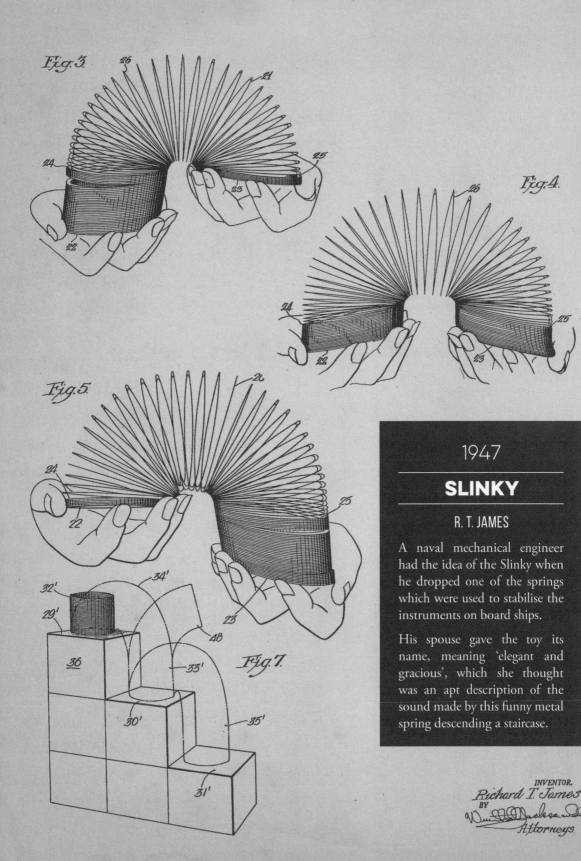

Fig.3. 26 21 25 24 23 22

Fig.4. 26 24 25 22 23

Fig.5. 21 24 25 22 23

32' 34' 29' 48 36 30' 33' *Fig.7.* 35' 31'

1947

SLINKY

R. T. JAMES

A naval mechanical engineer had the idea of the Slinky when he dropped one of the springs which were used to stabilise the instruments on board ships.

His spouse gave the toy its name, meaning 'elegant and gracious', which she thought was an apt description of the sound made by this funny metal spring descending a staircase.

INVENTOR.
Richard T. James
BY
Wm. Jackson
Attorneys.

19

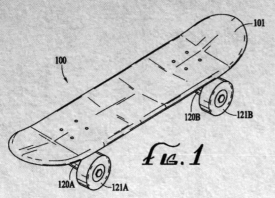

fig.1

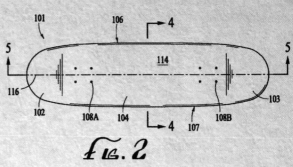

fig.2

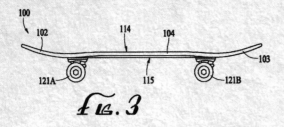

fig.3

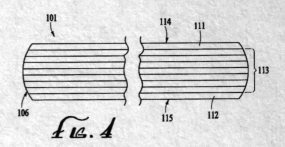

fig.4

2003

SKATEBOARD

S. GORDON

The first, rudimentary skateboards were scooters made from old crates with wheels on the bottom, during the impoverished days of the Great Depression in 1930s America. The wheels of roller skates, which had been patented in England in 1876, were gradually attached to boards.

The first manufactured models appeared in the 1950s under the brand Humco. Nicknamed 'roll-surfs' or 'sidewalk surfboards', they met with immediate success.

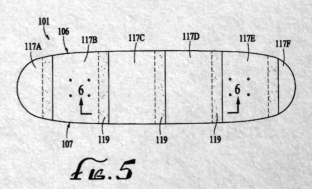

fig.5

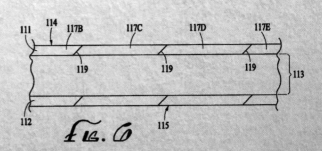

fig.6

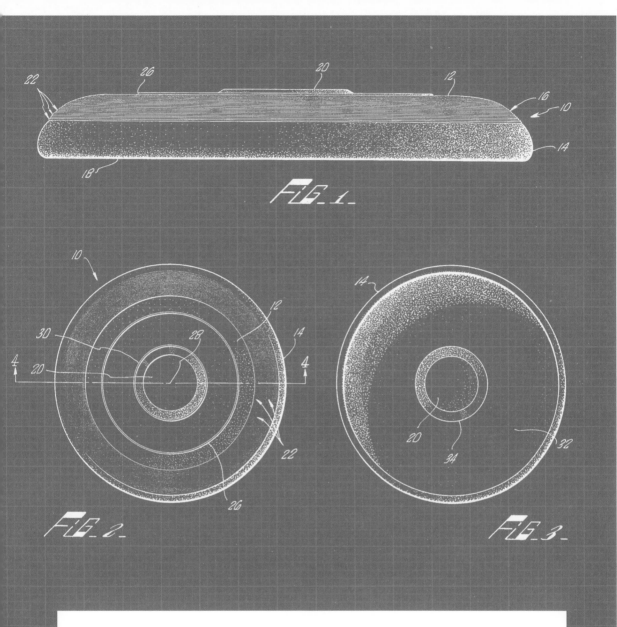

1967

FRISBEE

E. E. HEADRICK

In the 1940s in the USA, students at Yale University amused themselves with round cake tins made by the Frisbie Pie Company. Seeing them, Frederick Morrison had the idea of creating a Bakelite disc for throwing, naming it the Flying Saucer.

After the re-purchasing of the patent in 1957, it was renamed the Pluto Platter, then one year later the Frisbee, in homage to its origins.

The world record for the furthest throw, held by the German Simon Lizotte, is 263.2 metres.

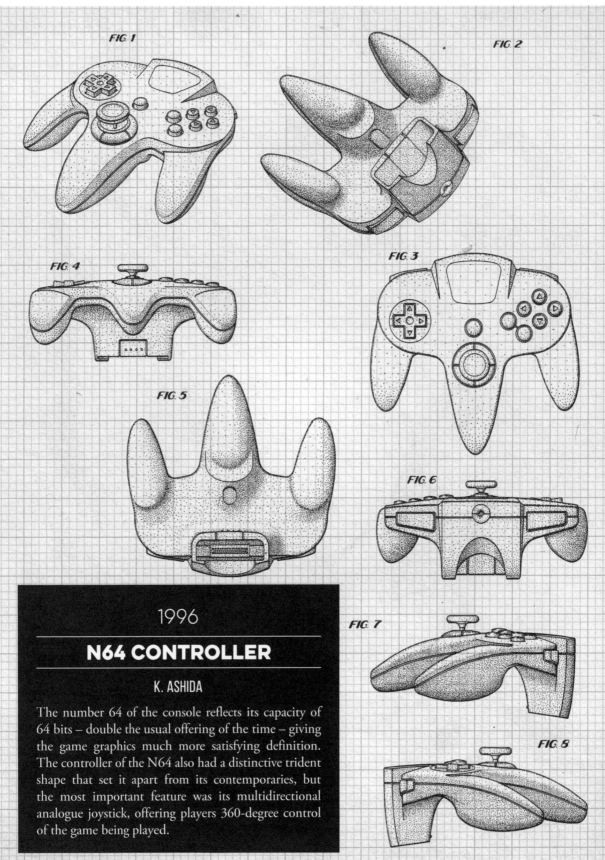

FIG. 1

FIG. 2

FIG. 4

FIG. 3

FIG. 5

FIG. 6

FIG. 7

FIG. 8

1996

N64 CONTROLLER

K. ASHIDA

The number 64 of the console reflects its capacity of 64 bits – double the usual offering of the time – giving the game graphics much more satisfying definition. The controller of the N64 also had a distinctive trident shape that set it apart from its contemporaries, but the most important feature was its multidirectional analogue joystick, offering players 360-degree control of the game being played.

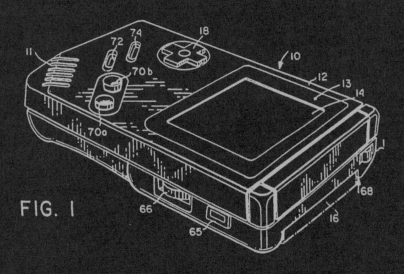

FIG. 1

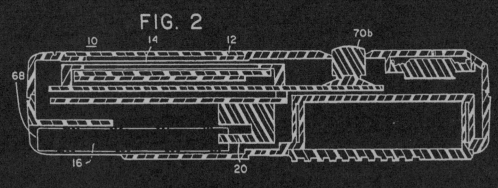

FIG. 2

FIG. 3

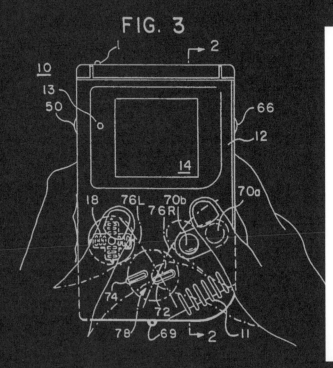

1993

GAME BOY

G. YOKOI

Nintendo's Game Boy was the creation of Gunpei Yokoi, who was working as a maintenance man for the company. Less powerful than some consoles of the time, it met with dazzling success thanks to its lower price and being bundled with the popular puzzle game *Tetris*. First released in Japan in 1989, its entire stock of 300,000 units sold out within two weeks. The Game Boy and its successor, the Game Boy Color, sold over 100 million units worldwide before the Game Boy Advance appeared in 2001.

Fig.1.

Fig.2.

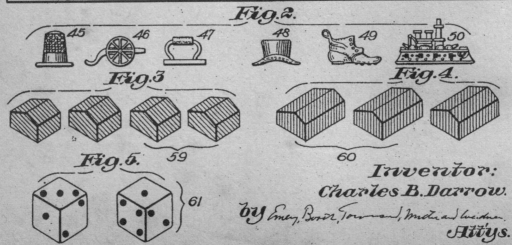

Fig.3.

Fig.4.

Fig.5.

Inventor:
Charles B. Darrow.

by Emey, Booth, Townsend, India and Weidner.
Attys.

24

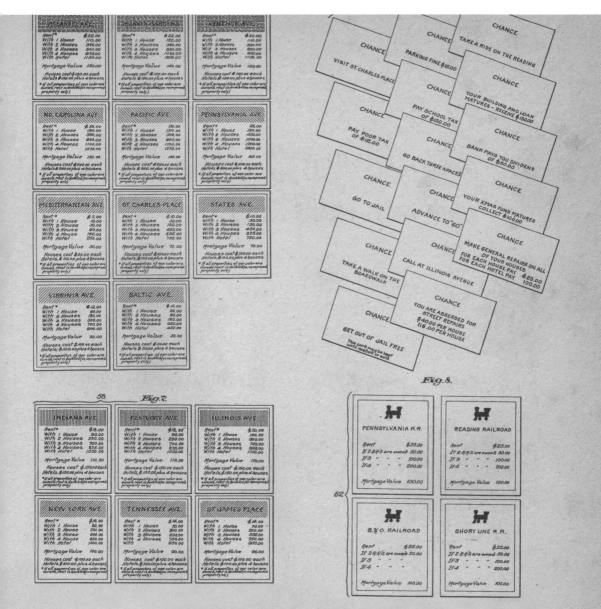

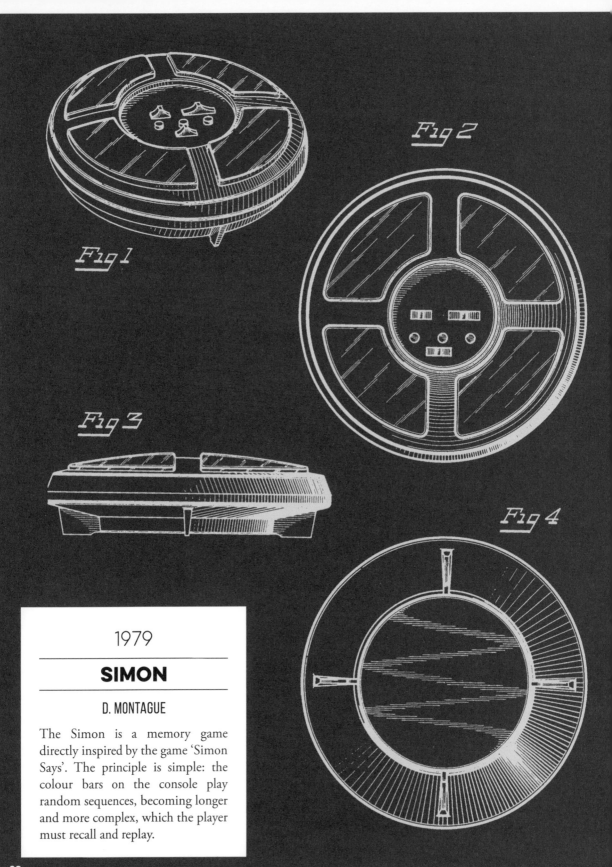

Fig 1

Fig 2

Fig 3

Fig 4

1979

SIMON

D. MONTAGUE

The Simon is a memory game directly inspired by the game 'Simon Says'. The principle is simple: the colour bars on the console play random sequences, becoming longer and more complex, which the player must recall and replay.

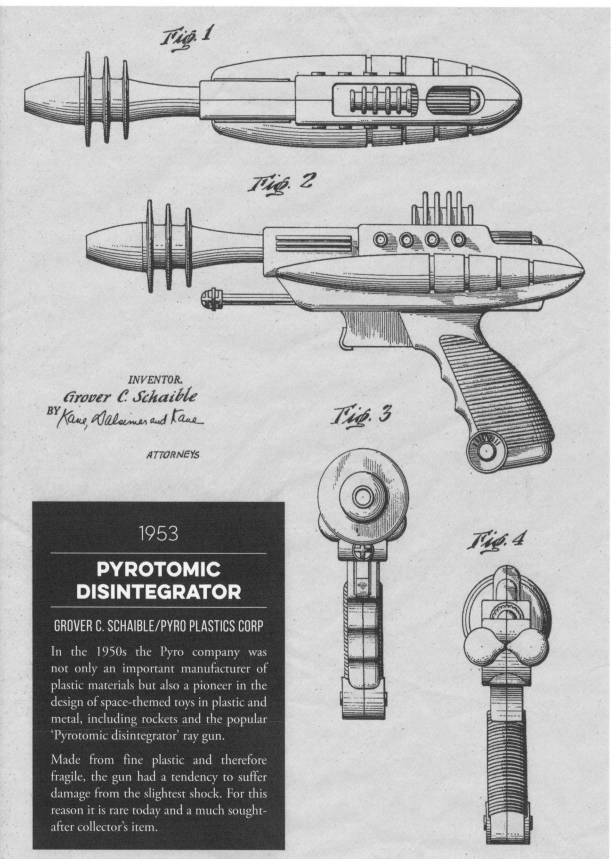

Fig. 1

Fig. 2

INVENTOR.

Grover C. Schaible

BY *Kane, Dalsimer and Kane*

ATTORNEYS

Fig. 3

Fig. 4

1953

PYROTOMIC DISINTEGRATOR

GROVER C. SCHAIBLE/PYRO PLASTICS CORP

In the 1950s the Pyro company was not only an important manufacturer of plastic materials but also a pioneer in the design of space-themed toys in plastic and metal, including rockets and the popular 'Pyrotomic disintegrator' ray gun.

Made from fine plastic and therefore fragile, the gun had a tendency to suffer damage from the slightest shock. For this reason it is rare today and a much sought-after collector's item.

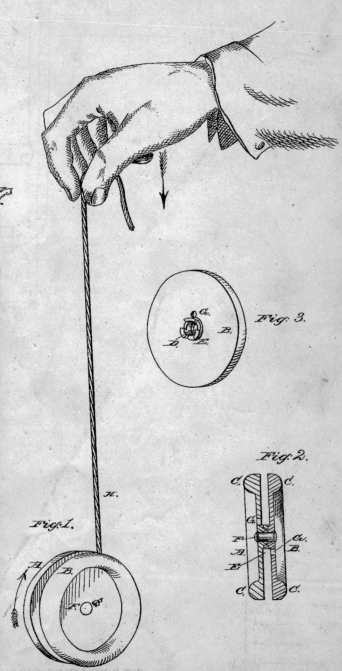

Haven & Hettrich,

Whirligig,

Nº 59,745.   *Patented Nov. 20, 1866.*

Witnesses:

Frank Millward
James H. Leyman.

Inventor:
James L. Haven
Charles Hettrich
By Knight Bros
Attys.

Fig. 3.

Fig. 2.

Fig. 1.

1866

YO-YO

J. L. HAVEN

Requiring skill and patience, the yo-yo is one of the oldest toys in the world, along with the spinning top. While it may have originated in China, there are also yo-yos depicted on ancient Greek pottery and the walls of Egyptian temples. In the eighteenth century it travelled from India to Europe.

The name 'yo-yo' probably came from the Philippines. But it was an American, Donald Duncan, who commercialised the toy on a large scale under its current name, registered in 1930.

SPORT AND LEISURE

In ancient times, the athletes of the Olympic Games were considered true heroes, like rock stars today, or certain legendary footballers and cyclists. But consider this: what would Jimi Hendrix have been without his guitar, Pelé without his football, Eddy Merckx without his bicycle?

In this chapter you will discover the patents of inventions which made their feats, their success, even their very careers possible. It's also thanks to objects such as these that we are able to entertain ourselves every day, while developing our physical and mental capacities.

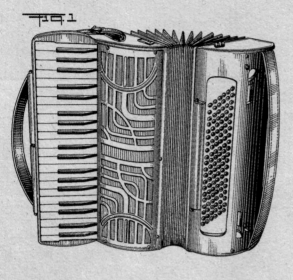

Fig. 1

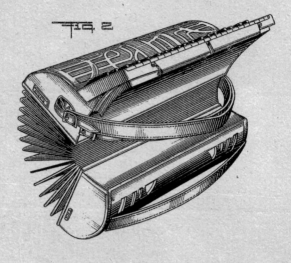

Fig. 2

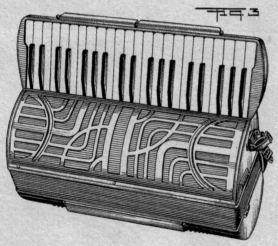

Fig. 3

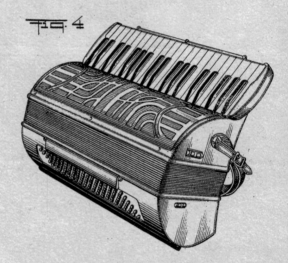

Fig. 4

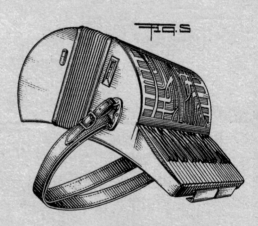

Fig. 5

1938

ACCORDION

J. VASSOS

Known also as a piano with straps, poor man's piano, thumbtack bellows, music box, squeezebox, harmonica and concertina, the accordion – which has over 50 names worldwide – was, it appears, the first patent registered in Vienna by a maker of organs and pianos, Cyrill Demian, in 1829.

The patent signed by John Vassos, resident of Connecticut, concerns the design of special ornaments.

INVENTOR.

BY John Vassos

Mock + Blum

ATTORNEYS

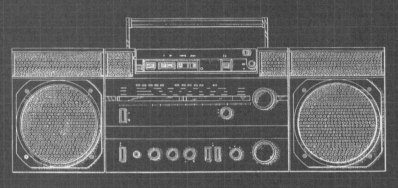

FIG. 1

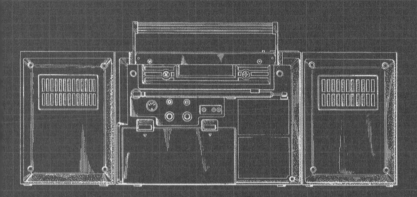

FIG. 5

1985

GHETTO BLASTER

V. S. SUJANANI

Cult radio cassette of the 1970s and 1980s, the ghetto blaster is known for its oversized body, powerful sound and portability. Invented in 1969 by the company Philips, it very quickly gained popularity with young music fans across the world, particularly in America, where it became an icon of hip hop culture.

The patent of Vinod S. Sujanani put in place an original system allowing the separation of the two speakers from the central block, in order to give a stereo effect.

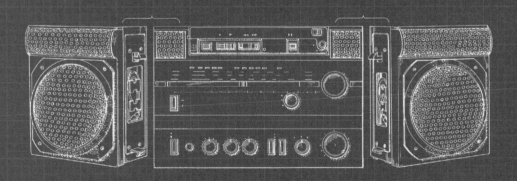

FIG. 9

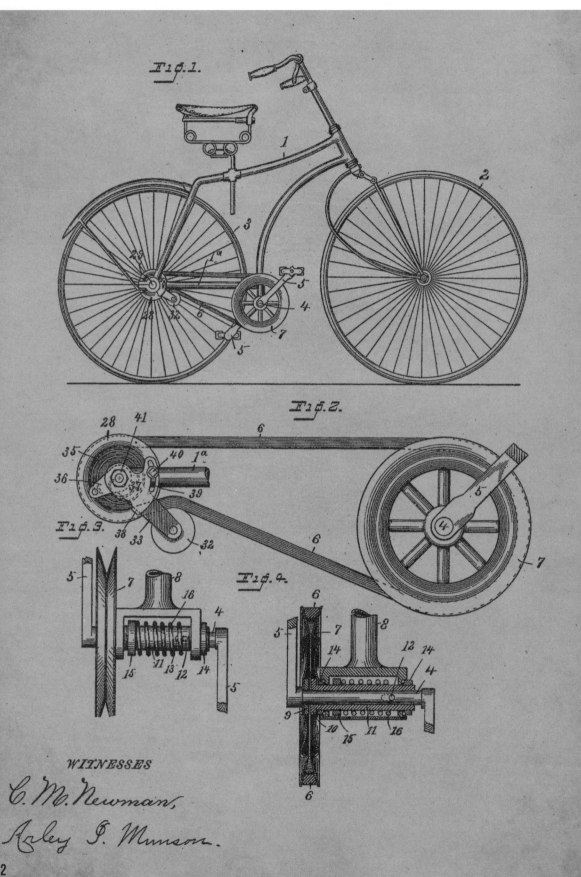

Fig.1.

Fig.2.

Fig.3.

Fig.4.

WITNESSES

C. M. Newman,
Ashley S. Munson.

32

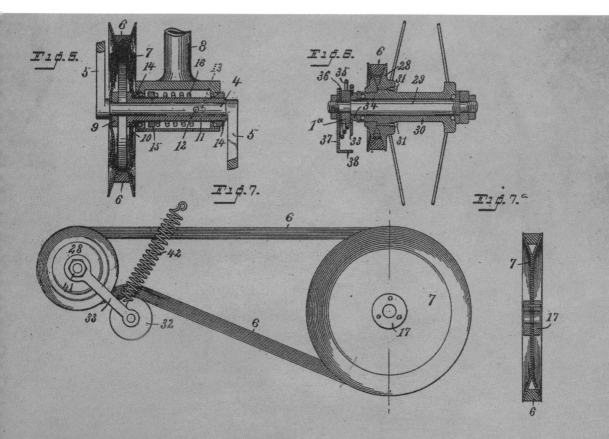

Fɪɢ.5.
Fɪɢ.6.
Fɪɢ.7.
Fɪɢ.7.ᵃ

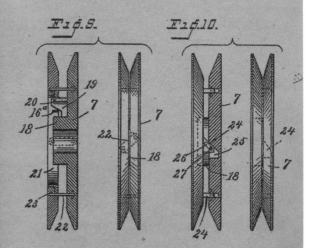

Fɪɢ.9.
Fɪɢ.10.

1890

BICYCLE GEARS

C. D. RICE

In April 1890 the American Charles D. Rice patented the first self-adjusting belt drive for a bicycle, to enable the rider to 'obtain a much higher rate of speed upon firm level ground than has heretofore been possible' and 'apply much greater power to hill climbing and riding upon heavy roads'.

The modern derailleur system of shifting or derailing the chain from one sprocket to the next emerged in the early twentieth century. A functional two-speed version was made by the French cycling pioneer Paul de Vivie in 1905.

INVENTOR
Charles D. Rice
By
H. M. Wooster
Atty.

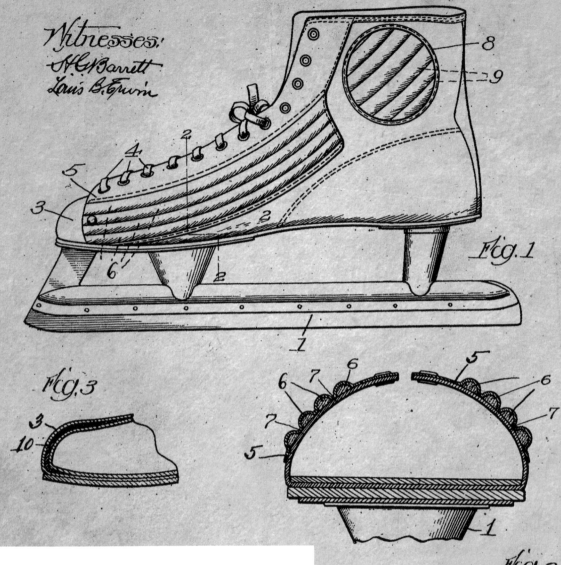

8
9
4
2
5
3
6
2
2
Fig.1
1
1

Fig.3
3
10

6
7
6
5
6
7
7
5
7
1
Fig.2

1914

ICE SKATES

N. JOHNSON

From the seventeenth century, ice skating became a popular pastime in British and Scandinavian high society, leading to the creation of shoes equipped with steel blades.

The 1914 patent of Nestor Johnson, resident of Chicago, is for a new type of skate developed for ice hockey players, which notably improved the skater's speed and offered greater stability on the ice.

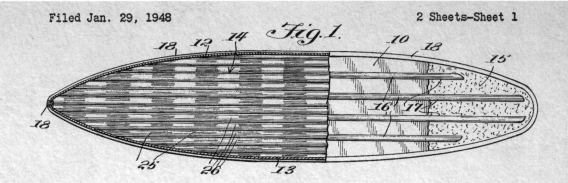

Fig. 1.

Fig. 2.

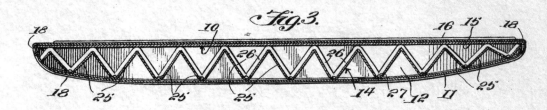

Fig. 3.

Fig. 4.

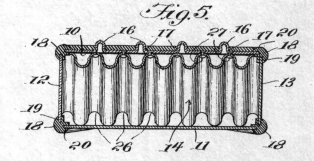

Fig. 5.

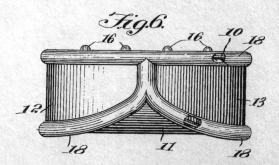

Fig. 6.

1950

SURFBOARD

G. B. D. PARKER

The oldest board to have survived was found in the tomb of a Hawaiian tribal chief. Discovered in 1905, it was made from coconut tree bark and was preserved in perfect condition.

Following the model of George B. D. Parker's patent, certain types of board began to distinguish themselves by having a hollow shell, offering better stability on the waves.

INVENTOR.

George B. D. Parker;

BY *Victor J. Evans & Co.*

ATTORNEYS

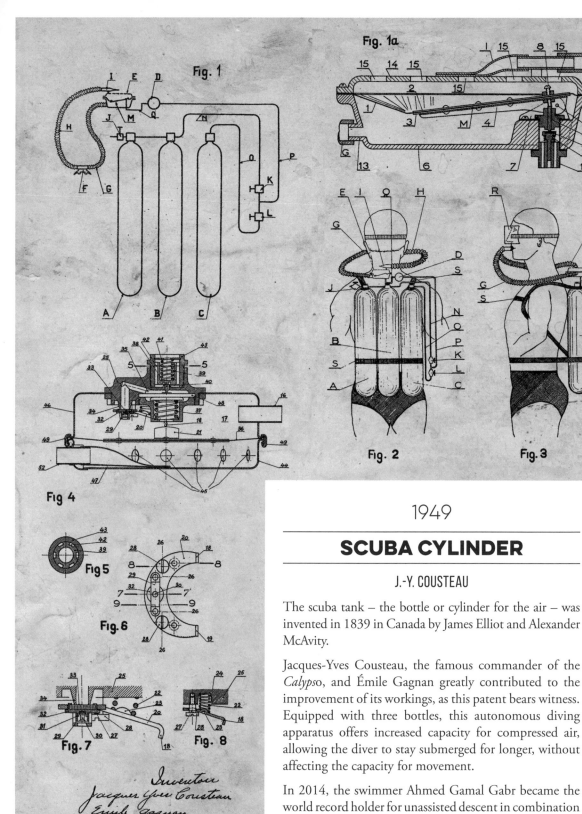

Fig. 1a

Fig. 1

Fig. 2

Fig. 3

Fig 4

Fig 5

Fig. 6

Fig. 7

Fig. 8

Inventors
Jacques Yves Cousteau
Émile Gagnan
By Pennie Edmonds Morton Barrows
attys

1949

SCUBA CYLINDER

J.-Y. COUSTEAU

The scuba tank – the bottle or cylinder for the air – was invented in 1839 in Canada by James Elliot and Alexander McAvity.

Jacques-Yves Cousteau, the famous commander of the *Calypso*, and Émile Gagnan greatly contributed to the improvement of its workings, as this patent bears witness. Equipped with three bottles, this autonomous diving apparatus offers increased capacity for compressed air, allowing the diver to stay submerged for longer, without affecting the capacity for movement.

In 2014, the swimmer Ahmed Gamal Gabr became the world record holder for unassisted descent in combination with cylinders. It took him 12 minutes to descend to a depth of 332.35 metres, and 15 hours to come back up.

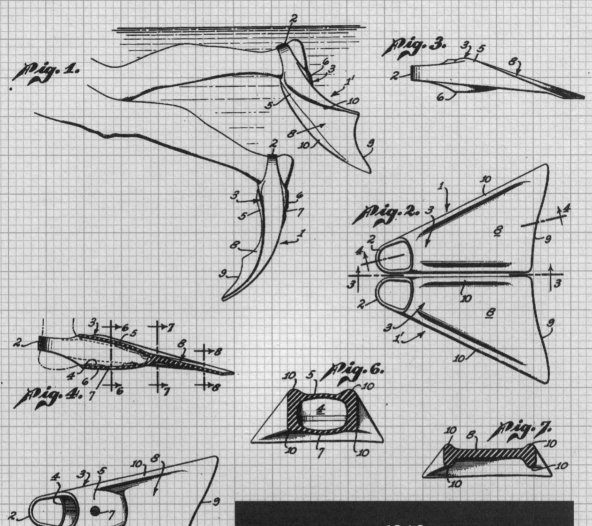

Fig. 1.

Fig. 3.

Fig. 2.

Fig. 4.

Fig. 6.

Fig. 7.

Fig. 5.

Fig. 8.

Inventor

OWEN P. CHURCHILL,

By

Attorney

1943

FLIPPERS

O. P. CHURCHILL

We owe the invention of the very first flippers to Louis de Corlieu, a lieutenant commander in the French Navy in 1914. In 1939, he decided to quit his post to begin production of his invention in his Paris apartment.

The patent of the American Owen P. Churchill is for a model intended for the general public, which could work for beginner divers as well as for fishing underwater.

The humble flipper is one of the most useful accessories for movement through water, allowing the diver to attain impressive speeds of up to 13 km/h.

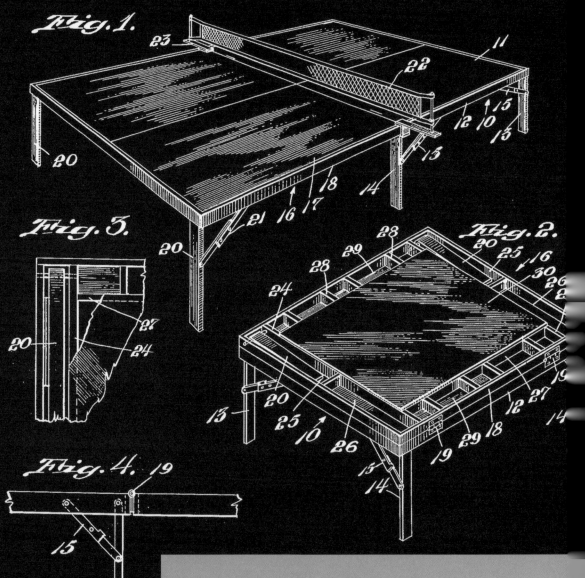

Fig. 1.

Fig. 3.

Fig. 2.

Fig. 4.

Fig. 5.

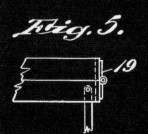

1952

PING-PONG TABLE

K. E. SUTTON

The most common story of the invention of table tennis is the following: over the course of a dinner, some noble Englishmen in the nineteenth century were discussing tennis and wanted to demonstrate some techniques on the table. To do this, they used a champagne cork as a ball and cigar boxes for rackets, lining up books to serve as the net.

In 1952, Kenneth E. Sutton created a folding ping-pong table, with storage boxes on the reverse that could hold packs of cards for poker, blackjack and so on.

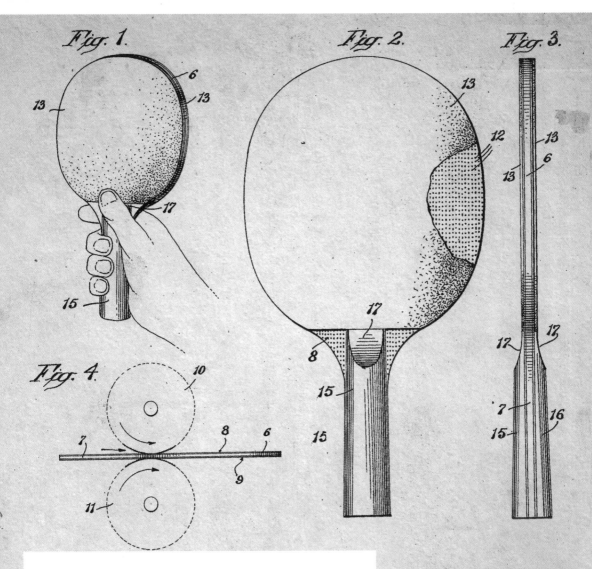

Fig. 1. **Fig. 2.** **Fig. 3.**

Fig. 4.

1935

PING-PONG BAT

G. H. PERRYMAN

George H. Perryman, a native of New Jersey, is considered the creator of the modern table-tennis bat, for which this patent was registered in 1935. Perryman's innovations marked an improvement in the bat's structure to make it more durable and comfortable to handle.

In 2014, the longest-ever ping-pong exchange was recorded using this type of bat: a rally of 8 hours, 40 minutes and 5 seconds, and 32,000 strokes, exchanged between Peter Ives and his son Daniel. Thanks to their feat, they collected close to £1,400 to help fund research into combating prostate cancer.

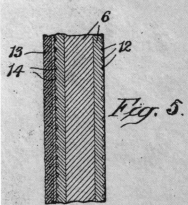

Fig. 5.

INVENTORS
George H. Perryman
Andrew J. MaCutchan
BY
Harry Radzinsky
ATTORNEY

Fig.1.

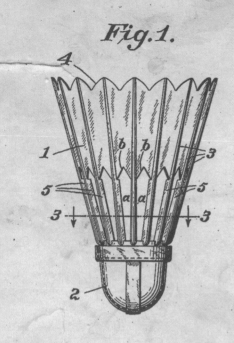

Fig.2.

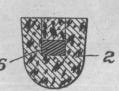

Fig.3.

Fig.4.

Fig.5.

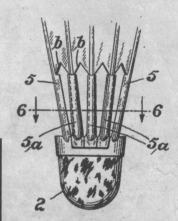

1927

SHUTTLECOCK

A. T. SAUNDERS

In 1873, at Badminton House in Gloucestershire, the Duke of Beaufort and several guests were forced by a storm to halt their outdoor ball game. It is uncertain whether the game was invented then, when two officers proposed to continue the game inside, replacing the ball with a champagne cork adorned with feathers, or whether the duke had brought the game back from India. In any case, the sport was thereafter known as badminton.

In 1927, the American Addison T. Saunders registered the patent for a new type of 'shuttle', equipped with a cork bottom and a feather-style 'skirt' which improved the trajectory. This type of model is still used in competitions today.

Fig.6.

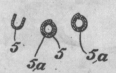

Inventor:

Addison T. Saunders,

by Spear, Middleton, Donaldson & Hall Attys.

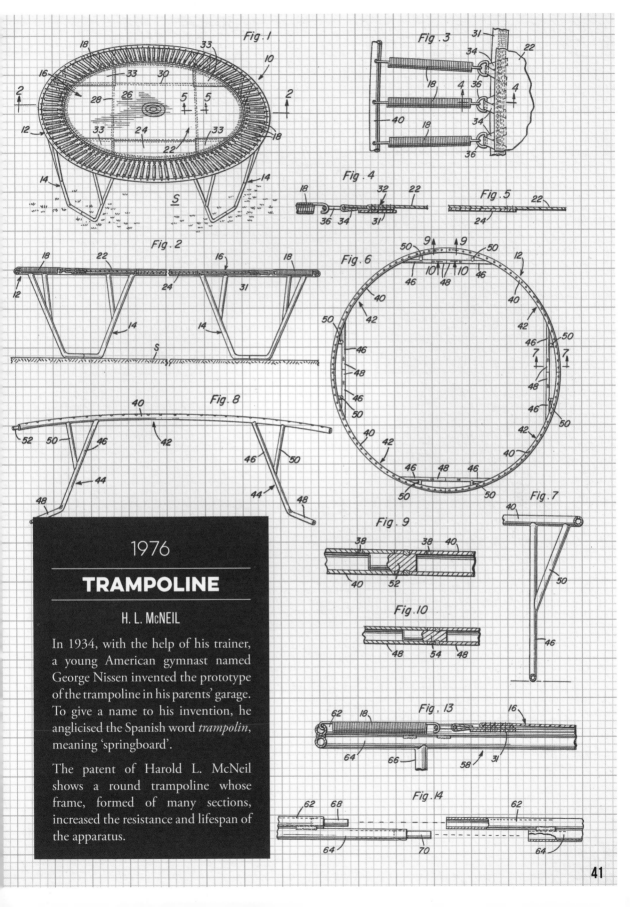

Fig. 1

Fig. 3

Fig. 4

Fig. 5

Fig. 2

Fig. 6

Fig. 8

Fig. 7

Fig. 9

Fig. 10

Fig. 13

Fig. 14

1976

TRAMPOLINE

H. L. McNEIL

In 1934, with the help of his trainer, a young American gymnast named George Nissen invented the prototype of the trampoline in his parents' garage. To give a name to his invention, he anglicised the Spanish word *trampolin*, meaning 'springboard'.

The patent of Harold L. McNeil shows a round trampoline whose frame, formed of many sections, increased the resistance and lifespan of the apparatus.

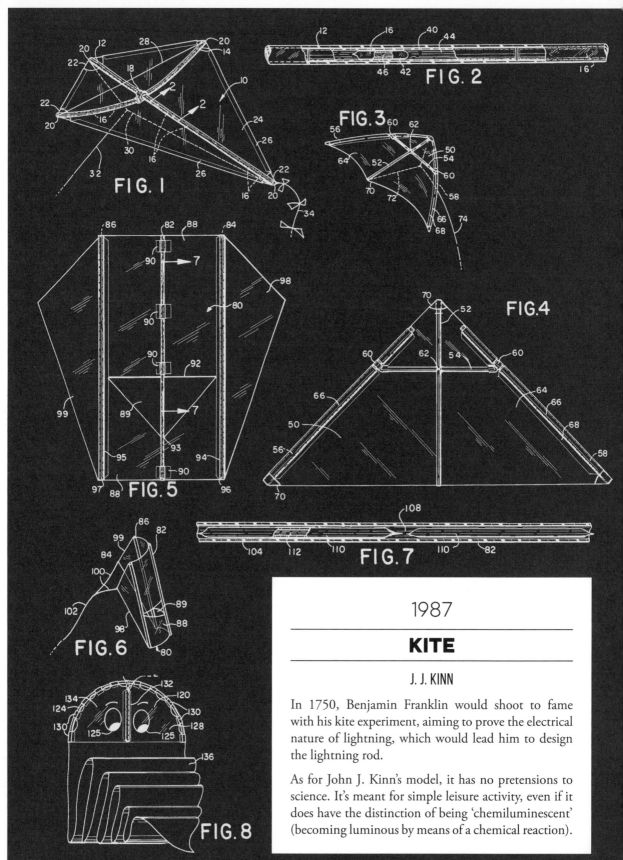

FIG. 1

FIG. 2

FIG. 3

FIG.4

FIG. 5

FIG. 6

FIG. 7

FIG.8

1987

KITE

J. J. KINN

In 1750, Benjamin Franklin would shoot to fame with his kite experiment, aiming to prove the electrical nature of lightning, which would lead him to design the lightning rod.

As for John J. Kinn's model, it has no pretensions to science. It's meant for simple leisure activity, even if it does have the distinction of being 'chemiluminescent' (becoming luminous by means of a chemical reaction).

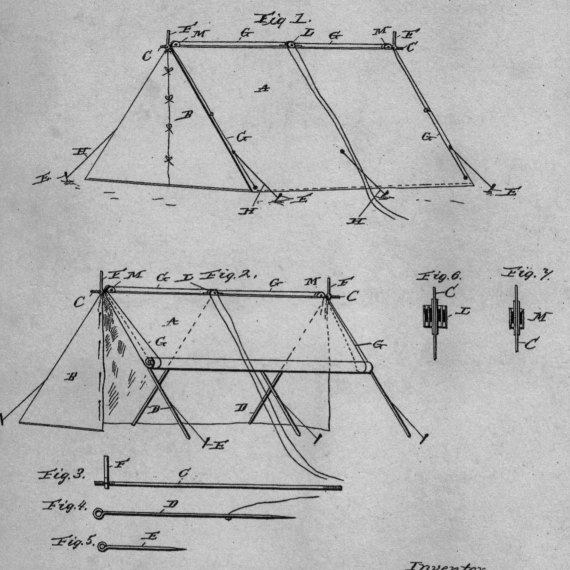

Fig. 1.

Fig. 2.

Fig. 3.

Fig. 4.

Fig. 5.

Fig. 6.

Fig. 7.

Inventor.

John J. Navo.

W.R. Stringfellow
Attorney

Witnesses:
C. H. Raeder.

James J. Shehy

1891

TENT

J. J. NAVO

John J. Navo, a resident of New Orleans, devised a pole tent which can open on the side with the aid of a sort of shutter.

The biggest tent in the world is the Khan Shatyr, pitched in Astana, the capital of Kazakhstan. With a height of 150 metres on a base of 140,000 m², it houses a giant entertainment complex.

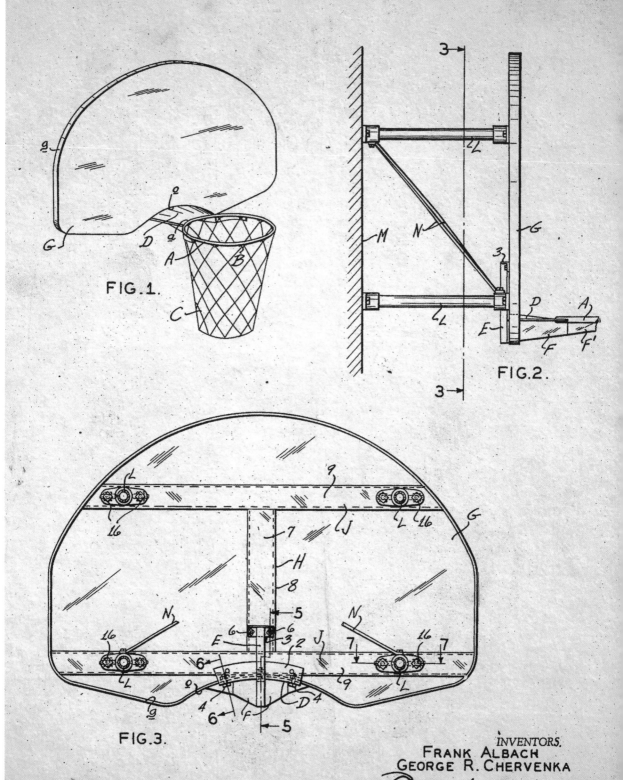

FIG.1.

FIG.2.

FIG.3.

INVENTORS.
FRANK ALBACH
GEORGE R. CHERVENKA

44

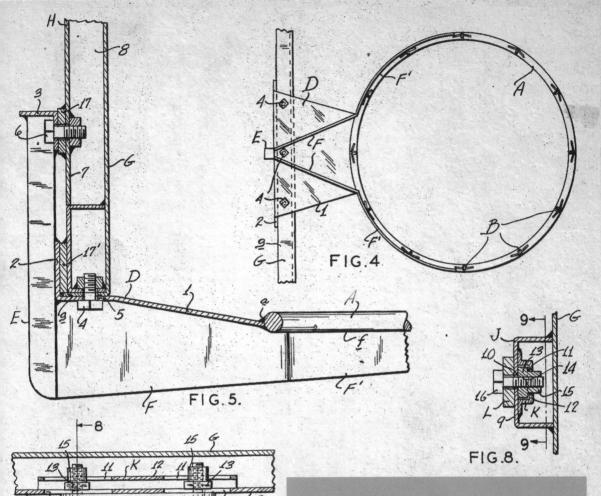

H

8

3

17.

6

G

7

17'

2

E

9   4   5

D

FIG. 5.

FIG. 4.

D

E

4

A

2

9

G

F'

F

A

F'

a

f

A

F'

9   G

J

10

13   11

14

16

15

L   9   K   12

FIG. 8.

8

15   15   G

13   11   K   12   11   13

9

10   10   J

16   16

8

L

FIG. 7.

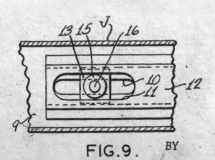

13 15 J   16

10

11

12

9

FIG. 9.   BY

1944

BASKETBALL HOOP

F. ALBACH

James Naismith, professor of sport education at a Canadian college, was looking for something to occupy his students during the winter between the football and baseball seasons.

Taking up the principle of an ancient Mayan ball game *(pok-ta-pok)*, he attached two peach baskets to a high railing of the gymnasium. This was 1891, and the idea of basketball was set in motion.

In 1894 the backboard was introduced, preventing the spectators from interfering with the play. The patent of American Frank Albach proposes a modern version of the basket, used by the NBA until the first league matches in 1949.

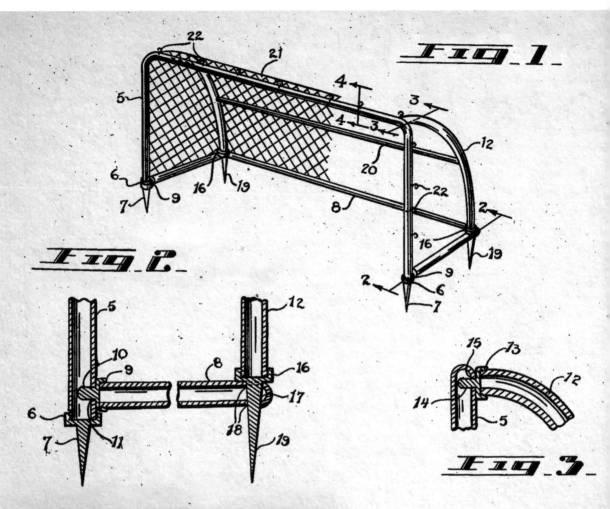

Fig. 1

Fig. 2

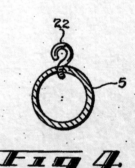

Fig. 3

Fig. 4

1937

FOOTBALL NET

S. OAKES

The first goal in professional football was scored by the Englishman Kenny Davenport on 6 September 1888. Contrary to legend, it wasn't an own goal by Gershom Cox for Aston Villa in their match that same afternoon against local rivals Wolves, but a 'nice shot' (according to the sports journalists present) from Davenport, the Bolton Wanderers right-winger, who scored two minutes after kick-off in the match against Derby County.

The patent of Canadian Stanley Oakes relates to football nets intended for amateurs, which could be easily installed and then taken down and put away.

INVENTOR
Stanley Oakes & Joseph Oakes

BY

ATTORNEY

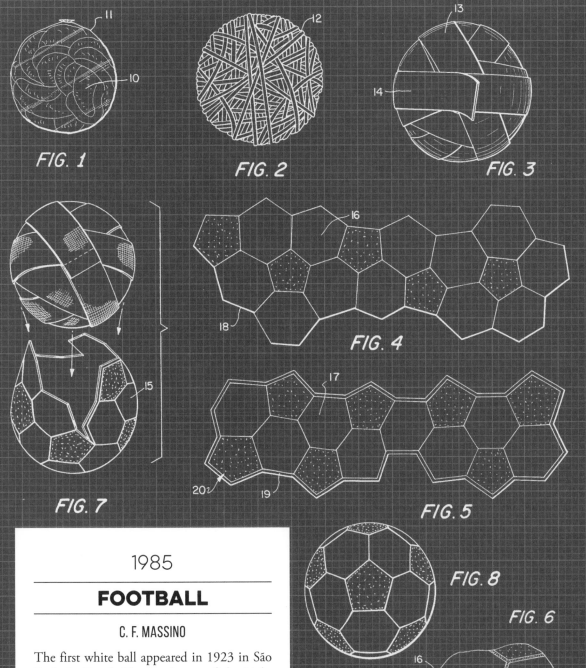

FIG. 1

FIG. 2

FIG. 3

FIG. 4

FIG. 5

FIG. 7

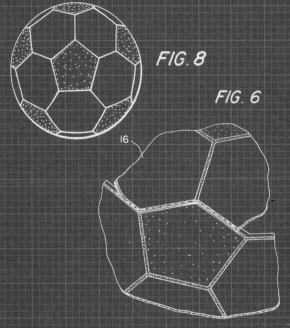

FIG. 8

FIG. 6

1985

FOOTBALL

C. F. MASSINO

The first white ball appeared in 1923 in São Paulo in Brazil. The employees of the Light and Power company played after work at their factory, and as the brown ball was difficult to see in the twilight, one of the players decided to paint it white.

American Chester Massino's ball (shown here) is designed to help young players and disabled people, is made from elastic fibre, which softens the feel without altering the aerodynamics.

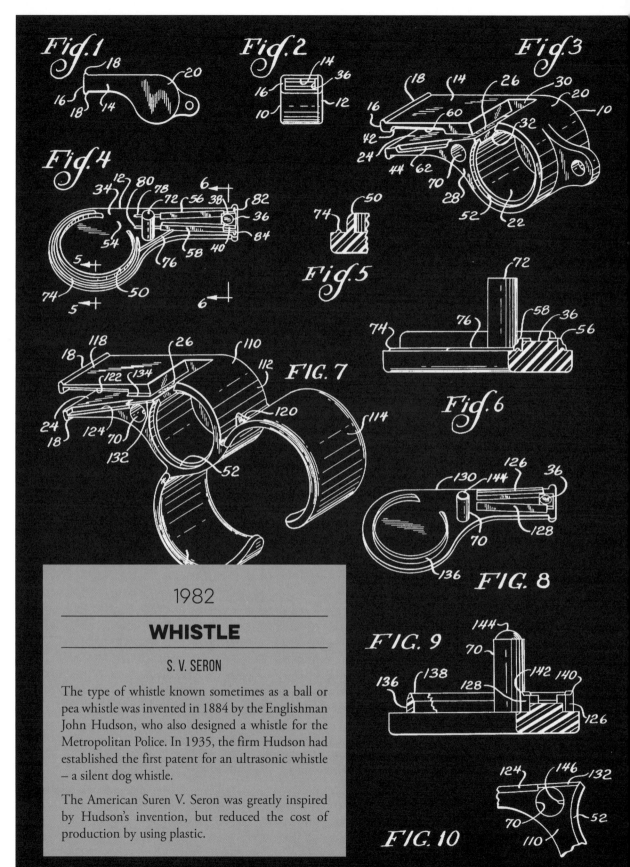

1982

WHISTLE

S. V. SERON

The type of whistle known sometimes as a ball or pea whistle was invented in 1884 by the Englishman John Hudson, who also designed a whistle for the Metropolitan Police. In 1935, the firm Hudson had established the first patent for an ultrasonic whistle – a silent dog whistle.

The American Suren V. Seron was greatly inspired by Hudson's invention, but reduced the cost of production by using plastic.

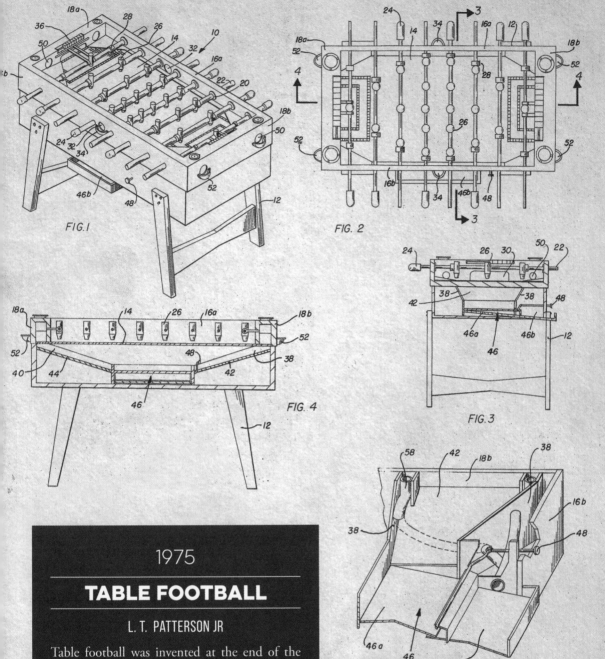

FIG.1

FIG. 2

FIG.3

FIG. 4

FIG. 5

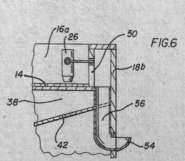

FIG.6

1975

TABLE FOOTBALL

L. T. PATTERSON JR

Table football was invented at the end of the nineteenth century – patents for games of this type date back to the 1890s – and became popular across Europe. The first UK patent for table football was filed by Tottenham Hotspur fan Harold Thornton in the 1920s.

Lawrence Patterson, an American soldier stationed in Germany in the 1960s, brought a table back to the USA in 1962 and trademarked the name 'foosball'.

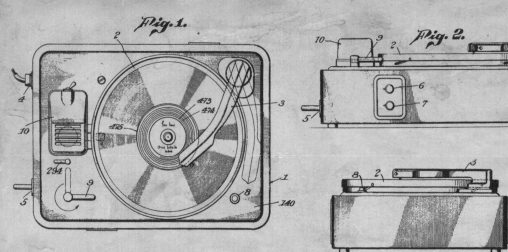

Fig. 1.

INVENTOR.
RALPH M. LIKE,
BY

ATTORNEY.

Fig. 2.

Fig. 3.

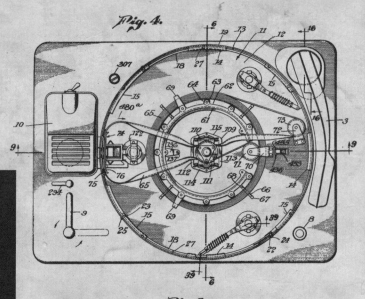

Fig. 4.

1950

PORTABLE TURNTABLE

R. M. LIKE

In 1946, Columbia Records marketed its first portable turntables, creating at the same time a huge craze for vinyl, which from then on could be heard pretty much everywhere.

It was around this time that the American inventor Ralph M. Like, to whom we owe this patent, attempted to commercialise a portable apparatus combining image and sound. It didn't catch on because the price was too expensive for consumers.

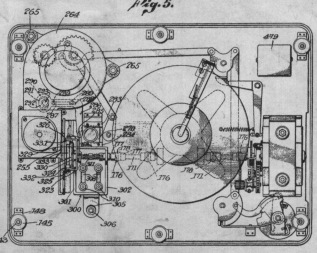

Fig. 5.

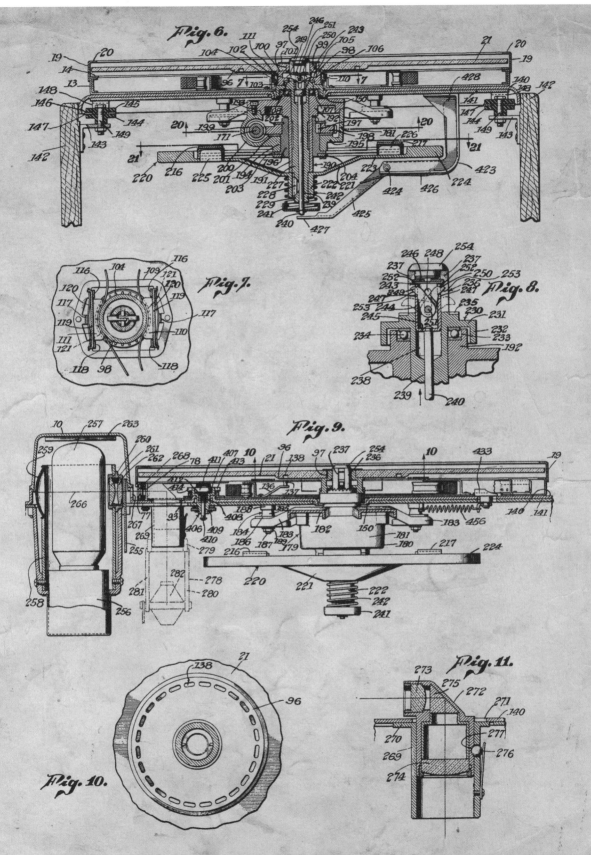

Fig. 6.

Fig. 7.

Fig. 8.

Fig. 9.

Fig. 10.

Fig. 11.

51

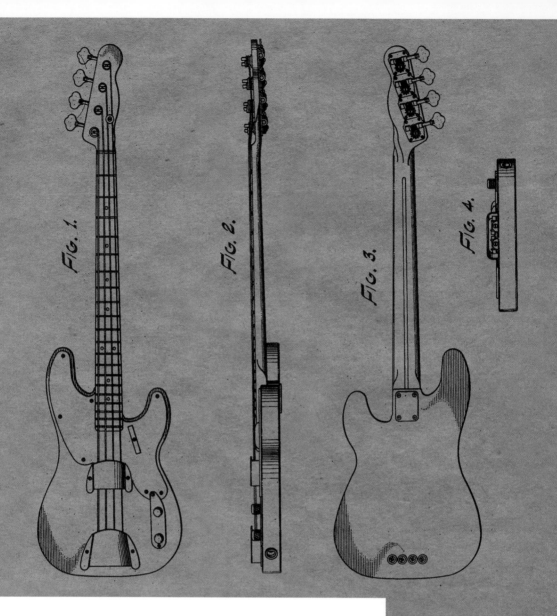

FIG. 1.   FIG. 2.   FIG. 3.   FIG. 4.

1953

P-BASS

C. L. FENDER

Clarence Leonidas Fender, known as Leo Fender, created his first guitar, Leo's Baby, at the age of sixteen. His iconic Telecaster appeared in 1951, followed by the Stratocaster in 1954.

Until Fender, the bass was a bulky upright instrument lacking volume. Leo designed a bass that could be played like a guitar through an amp.

The P-Bass, or Precision Bass, is the first patent he registered for an electric bass. A large majority of bass players still use it today.

INVENTOR.
CLARENCE L. FENDER
BY
Lyon & Lyon
ATTORNEYS

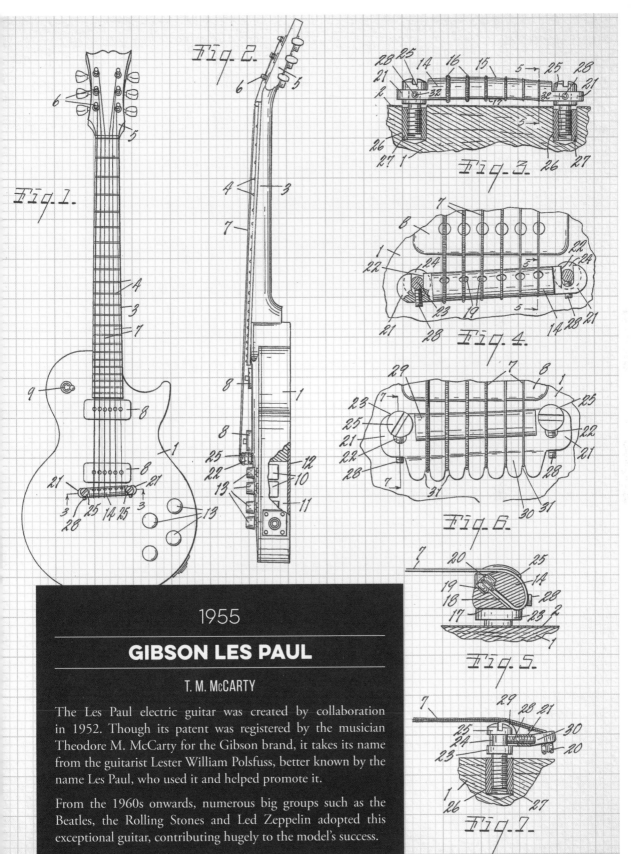

Fig. 1.
Fig. 2.
Fig. 3.
Fig. 4.
Fig. 5.
Fig. 6.
Fig. 7.

1955

GIBSON LES PAUL

T. M. McCARTY

The Les Paul electric guitar was created by collaboration in 1952. Though its patent was registered by the musician Theodore M. McCarty for the Gibson brand, it takes its name from the guitarist Lester William Polsfuss, better known by the name Les Paul, who used it and helped promote it.

From the 1960s onwards, numerous big groups such as the Beatles, the Rolling Stones and Led Zeppelin adopted this exceptional guitar, contributing hugely to the model's success.

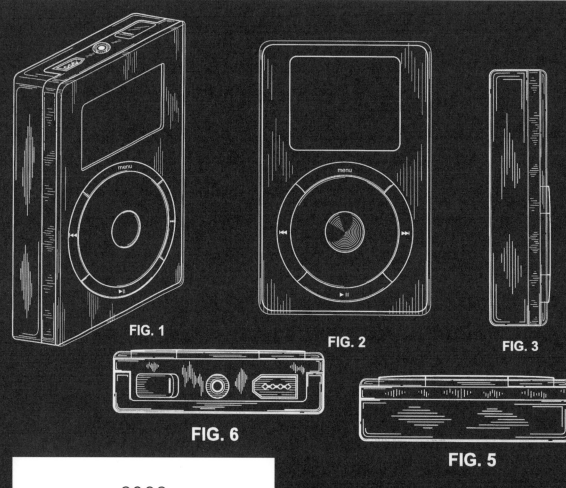

FIG. 1

FIG. 2

FIG. 3

FIG. 6

FIG. 5

2003

IPOD

B. K. ANDRE

In 2001, Apple launched its digital Walkman, the iPod, with immediate success. In 2010, the number of units sold, including all models, approached some 275 million.

Did you know that the famous 'i' at the start of the names of Apple products almost didn't happen? In the 1980s, Steve Jobs wanted to replace the name Macintosh with Macman. It was the Apple marketing team who suggested iMac, which would evoke the words 'internet', 'individual', 'imagination' and 'innovation'. If this suggestion hadn't been taken up, what Bartley K. Andre designed would have been called a Podman.

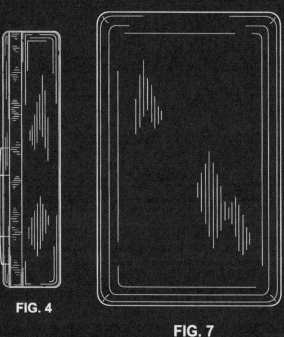

FIG. 4

FIG. 7

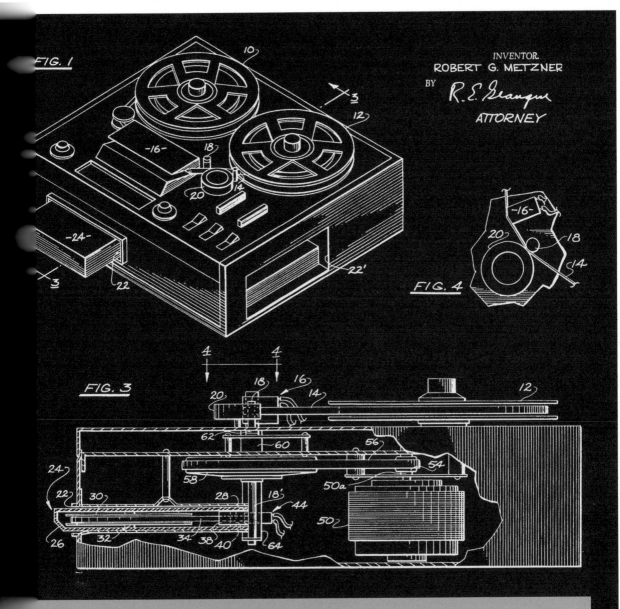

FIG. 1

INVENTOR.
ROBERT G. METZNER
BY
R. E. Granger
ATTORNEY

FIG. 4

FIG. 3

1968

CASSETTE RECORDER

R. G. METZNER

Using magnetic tape, reel-to-reel recorders had existed for several decades before Philips pioneered the compact cassette in the 1960s, making audio recording and listening far more accessible to the general public. Between the early 1970s and the early 2000s, the cassette was one of the two most common formats for listening to pre-recorded music – first alongside the LP record, then later the compact disc.

This patent from American entrepreneur and inventor Robert G. Metzner presented a device that included both a conventional reel-to-reel tape recorder and the means of playing a tape cassette, using a single drive mechanism.

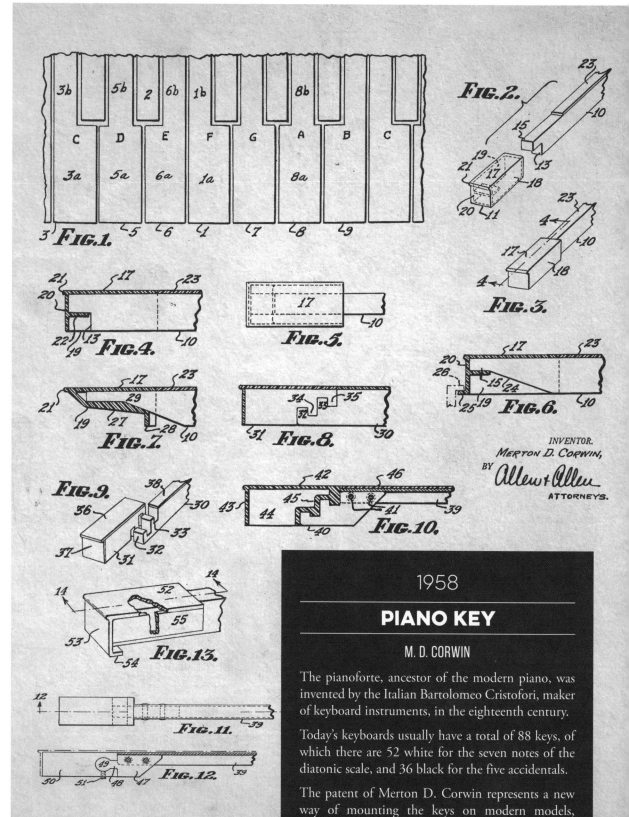

FIG.1.

FIG.2.

FIG.3.

FIG.4.

FIG.5.

FIG.6.

FIG.7.

FIG.8.

FIG.9.

FIG.10.

FIG.11.

FIG.12.

FIG.13.

INVENTOR.
MERTON D. CORWIN,
BY
Allen & Allen
ATTORNEYS.

1958

PIANO KEY

M. D. CORWIN

The pianoforte, ancestor of the modern piano, was invented by the Italian Bartolomeo Cristofori, maker of keyboard instruments, in the eighteenth century.

Today's keyboards usually have a total of 88 keys, of which there are 52 white for the seven notes of the diatonic scale, and 36 black for the five accidentals.

The patent of Merton D. Corwin represents a new way of mounting the keys on modern models, intended to simplify and cheapen the construction.

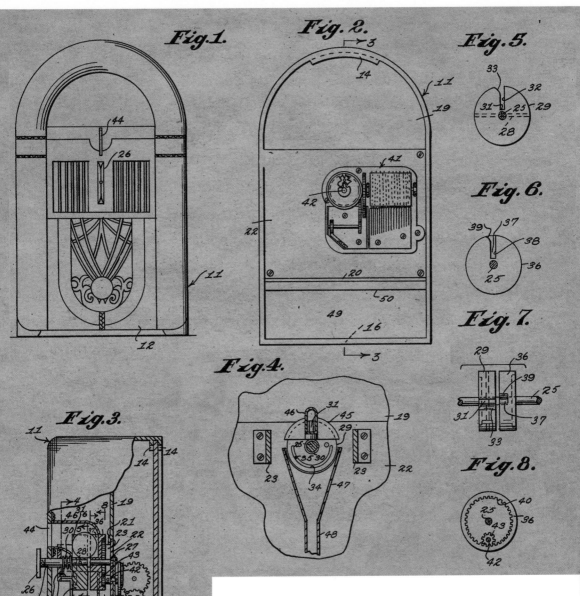

1950

JUKEBOX

R. H. EDWARDS

The term jukebox appeared a decade after its invention by the Frenchman Michel Bussoz, in 1920. The name made reference to the slang 'juke joints', which were bars where people danced, and 'juke bands', the groups that played in them.

The patent registered by the American Russell H. Edwards shows a model resembling those which adorned bars all over the USA between 1940 and 1980. During this period, almost two million units were produced and sold there.

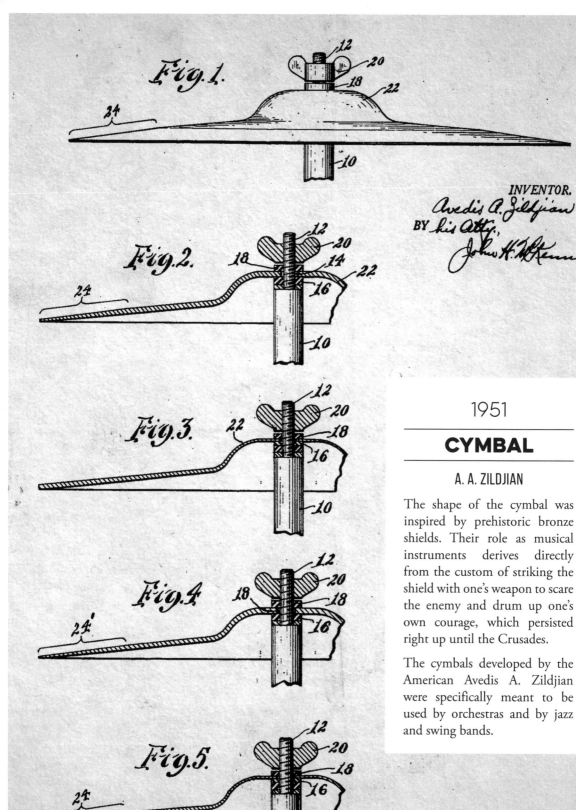

Fig. 1.

Fig. 2.

Fig. 3.

Fig. 4.

Fig. 5.

INVENTOR.

Avedis A. Zildjian

BY his Atty,

John H. McKenna

1951

CYMBAL

A. A. ZILDJIAN

The shape of the cymbal was inspired by prehistoric bronze shields. Their role as musical instruments derives directly from the custom of striking the shield with one's weapon to scare the enemy and drum up one's own courage, which persisted right up until the Crusades.

The cymbals developed by the American Avedis A. Zildjian were specifically meant to be used by orchestras and by jazz and swing bands.

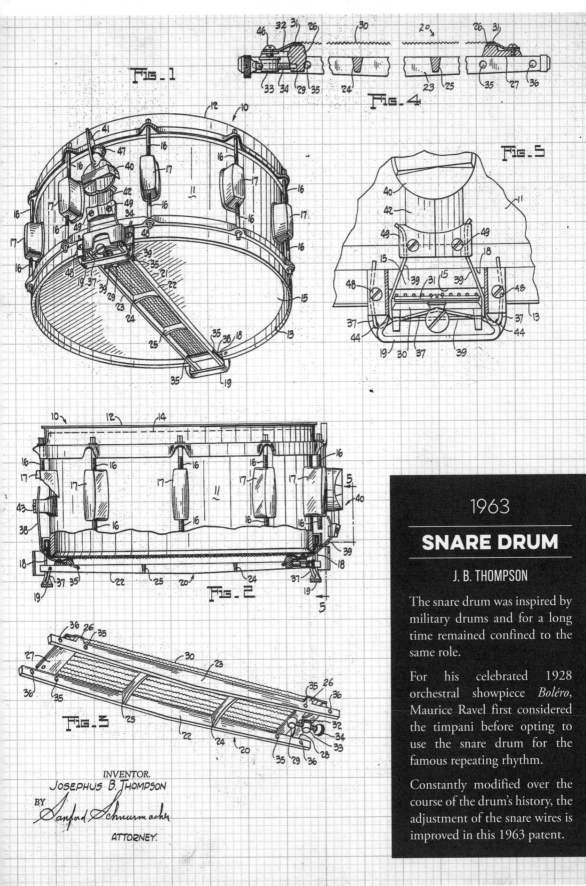

FIG. 1

FIG. 4

FIG. 5

FIG. 2

FIG. 3

INVENTOR.
JOSEPHUS B. THOMPSON

BY
Sanford Schnurmacher

ATTORNEY.

1963

SNARE DRUM

J. B. THOMPSON

The snare drum was inspired by military drums and for a long time remained confined to the same role.

For his celebrated 1928 orchestral showpiece *Boléro*, Maurice Ravel first considered the timpani before opting to use the snare drum for the famous repeating rhythm.

Constantly modified over the course of the drum's history, the adjustment of the snare wires is improved in this 1963 patent.

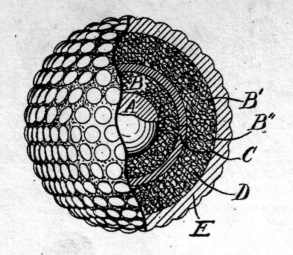

B'
B"
C
D
E

Witnesses
Herbert J. Smith.
R. W. Pitman

Inventor:
Richard B. Cavanagh.
By his Attorney,
F. W. Barnacle,

Fig.1.

B

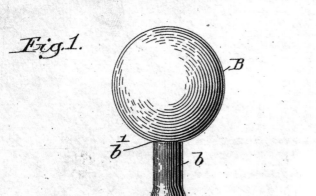

$\frac{1}{b}$

b

a^2

a

Fig.2.

b^1

$\frac{1}{a}$

b

x — x

a^2

a

Fig.3.

$\frac{1}{a}$
$\frac{1}{b}$

1902

MULTILAYER GOLF BALL

R. B. CAVANAGH

Initially, golf balls were made from leather stuffed with boiled feathers. The modern golf ball, developed in 1898 by an amateur named Coburn Haskell, made the game much more popular. It is characterised by the addition of several layers of rubber, as is evident on the patent developed by Richard B. Cavanagh. His improvements were intended to combine low price and durability.

Guinness World Records attribute the longest drive ever achieved in competition to Mike Austin, then aged 64 years. The ball travelled 471 metres across the Winterwood Golf Course in Nevada in 1974.

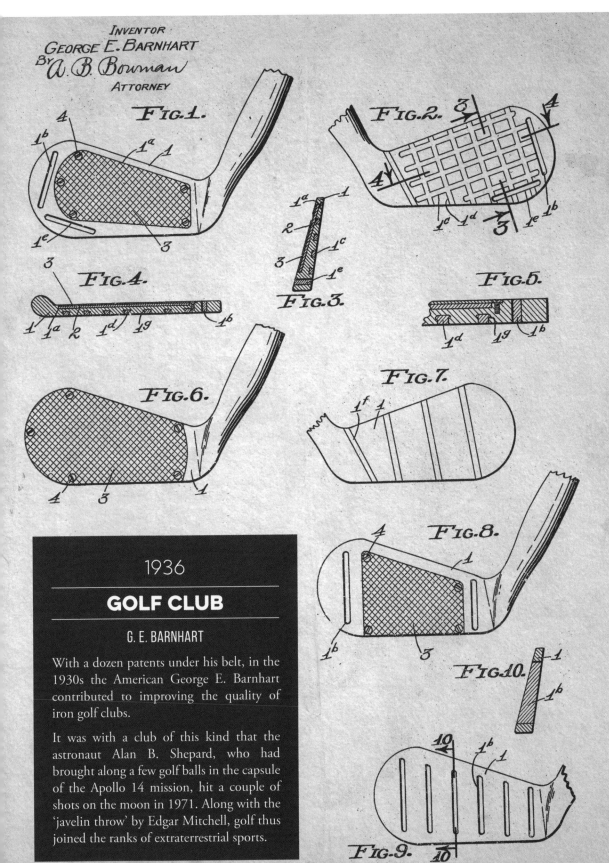

INVENTOR
GEORGE E. BARNHART
BY A. B. Bowman
ATTORNEY

FIG.1.

FIG.2.

FIG.3.

FIG.4.

FIG.5.

FIG.6.

FIG.7.

FIG.8.

FIG.9.

FIG.10.

1936

GOLF CLUB

G. E. BARNHART

With a dozen patents under his belt, in the 1930s the American George E. Barnhart contributed to improving the quality of iron golf clubs.

It was with a club of this kind that the astronaut Alan B. Shepard, who had brought along a few golf balls in the capsule of the Apollo 14 mission, hit a couple of shots on the moon in 1971. Along with the 'javelin throw' by Edgar Mitchell, golf thus joined the ranks of extraterrestrial sports.

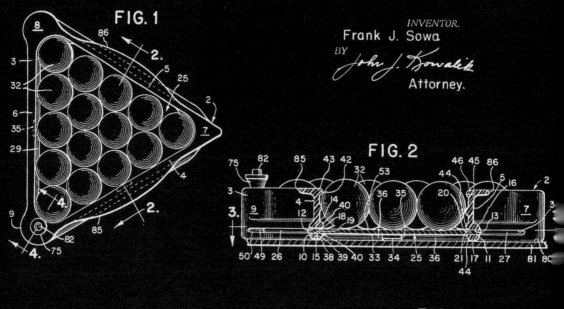

INVENTOR.
Frank J. Sowa
BY
John J. Kowalik
Attorney.

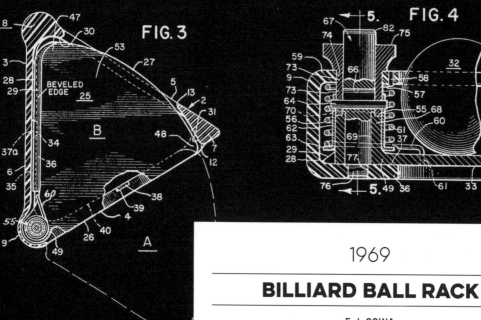

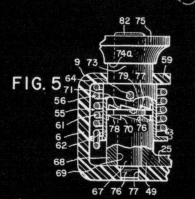

1969

BILLIARD BALL RACK

F. J. SOWA

For a long time it was claimed that the word 'billiards' came from its inventor, a London tailor called Bill. As the story goes, he played with three balls, which he pushed across his counter with his tailor's ruler – hence the name given to the game in 1560, Bill's Yard. It's a nice anecdote, but it's not true, as by then the game had already existed for three centuries, with the first tables appearing in 1469.

The patent of the American Frank J. Sowa concerns the rack allowing the player to place the balls in a triangle before the start of the game.

TOOLS AND UTENSILS

It's hard to imagine certain professions without the objects we automatically associate with them: the doctor's stethoscope, the fireman's hose, the barber's chair, the plumber's toolbox, and so on and so forth.

All of these accessories have evolved through the decades, becoming more and more practical, even more precise and increasingly easier to use. Here is a selection of these tools and utensils, accompanied by their patents.

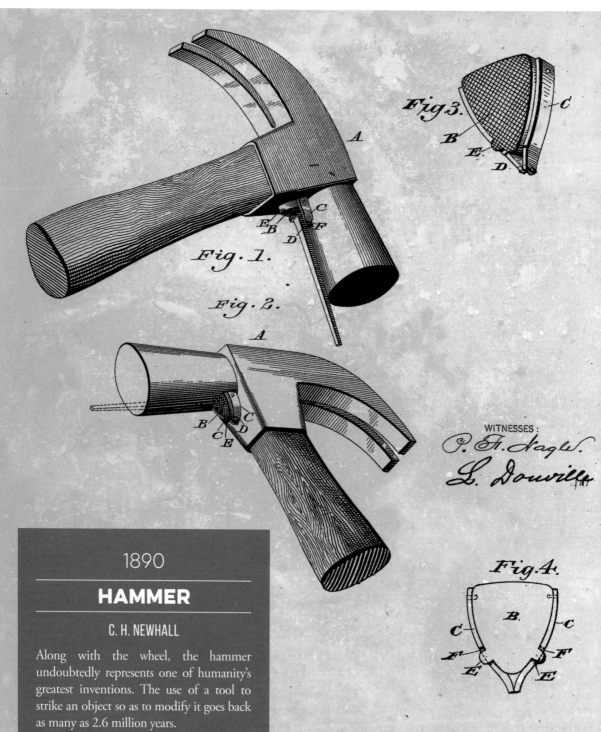

Fig. 3.

Fig. 1.

Fig. 2.

Fig. 4.

WITNESSES:
P. H. Nagle
L. Douville

INVENTOR:
Cebert H. Newhall
BY John A. Diedersheim
ATTORNEY.

1890

HAMMER

C. H. NEWHALL

Along with the wheel, the hammer undoubtedly represents one of humanity's greatest inventions. The use of a tool to strike an object so as to modify it goes back as many as 2.6 million years.

Today, there exist at least 19 types of hammer. The tool patented by Cebert H. Newhall has the particular characteristic of possessing two heads, of which one, equipped with 'jaws', allows you to hold the nail in place or remove it.

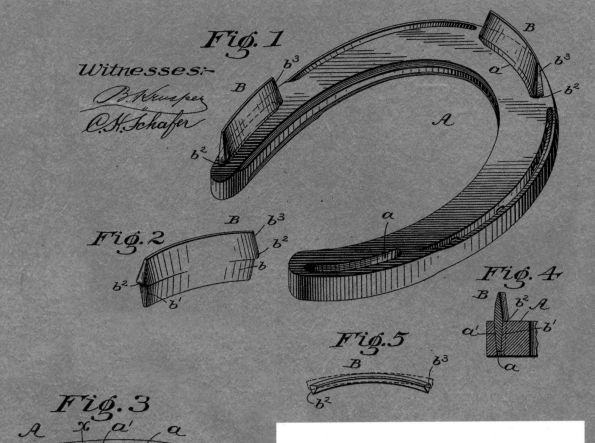

Fig. 1

Witnesses:-
B. Krusper
C. H. Schafer

Fig. 2

Fig. 3

Fig. 4

Fig. 5

Fig. 6

1898

HORSESHOE

P. AND J. P. HOPPESCH

The ancestor of the horseshoe was a Roman invention. The 'hipposandal' ('hippo' coming from the Greek for horse) was a piece of metalwork covering the hoof and held on by leather ties. The patent of Peter and John P. Hoppesch presents a more rigorous method of fixing it in place, which also made it easier to remove.

Legend has it that the Roman emperor Nero used to shoe his horses with silver, while his wife Poppaea shoed her mules with gold. Hence horseshoes became associated with good luck.

Inventors
Peter Hoppesch
John P. Hoppesch

By their Atty.

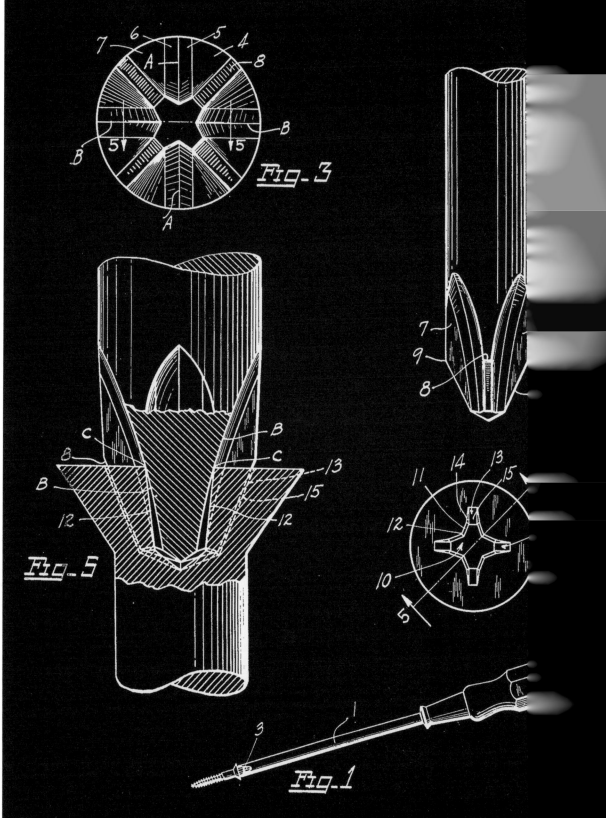

Fig-3

Fig-5

Fig-1

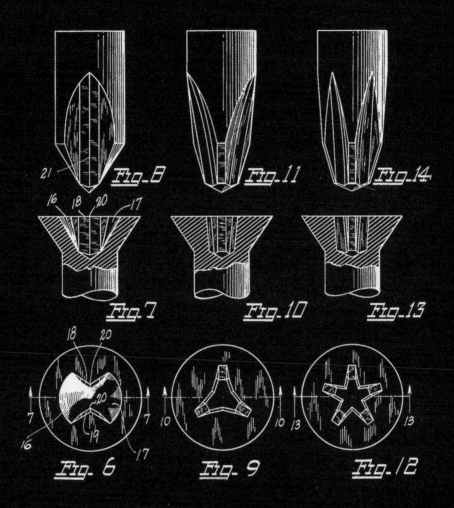

1936

SCREW AND SCREWDRIVER

H. F. PHILLIPS

In the 1930s, Henry F. Phillips developed a new form of screw, which diminished the risk of the screwdriver slipping, which was frequent with slotted screws. This first model of cruciform screw would be christened with the name of its inventor.

The sheets of this patent illustration show different phases of the operation of the screwdriver and various types of screw. Figure 4 notably shows the head of the famous 'Phillips' screw; figure 6, that of a 'Clutch' model; figure 9, a triangular head; figure 12, that of a 'Torx' screw with a star shape.

HENRY F. PHILLIPS.
INVENTOR.

BY *James D. Givnan*
ATTORNEYS.

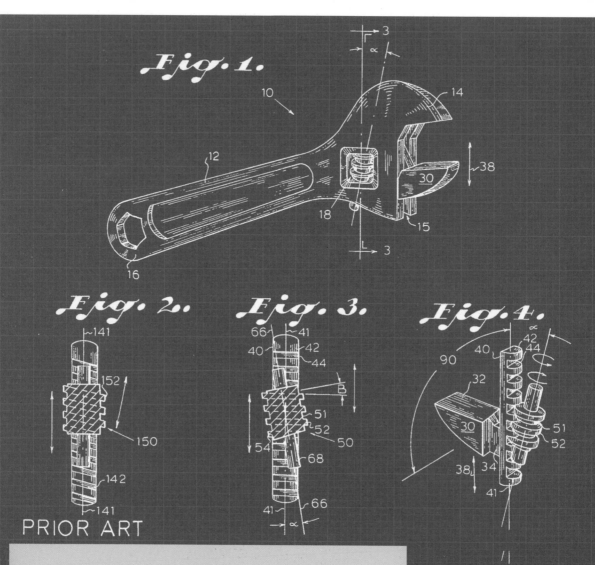

Fig. 1.

10 · 14 · 12 · 38 · 30 · 18 · 15 · 16 · α · 3

Fig. 2.

141 · 152 · 150 · 142 · 141

Fig. 3.

66 · 41 · 40 · 42 · 44 · B · 51 · 52 · 50 · 54 · 68 · 66 · 41 · α

Fig. 4.

α · 42 · 44 · 40 · 90 · 32 · 30 · 51 · 52 · 38 · 34 · 41

1977

ADJUSTABLE SPANNER

W. CAGLE

We owe the invention of the adjustable spanner to Johan Petter Johansson, a Swedish industrialist who registered around 118 patents over the course of his life.

In 1890, Hjort & Company acquired the rights to sell his inventions in order to create what would become the Bahco company, today a market leader in hand tools. Since its foundation, it is estimated that over 100 million units of the adjustable spanner have been sold by Bahco.

In 1977, William Cagle perfected the Johansson model and proposed a way of locking the moving jaw of the wrench by means of pressure, thus making for more precise usage.

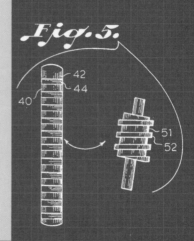

Fig. 5.

42 · 44 · 40 · 51 · 52

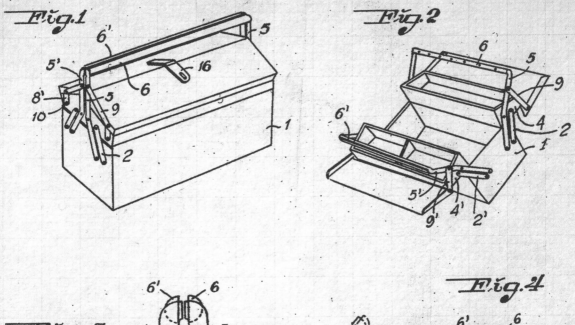

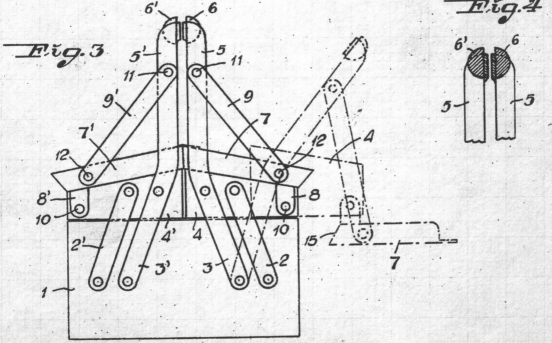

1944

TOOLBOX

E. LARSSON

This could be the first patent for a modern toolbox. The innovation of the Swede Eric Larsson aimed to give simultaneous access to both the main compartment and the adjoining receptacles by the simple act of opening the box.

Inventor:
Eric Larsson,
by Pierce & Scheffler,
Attorneys.

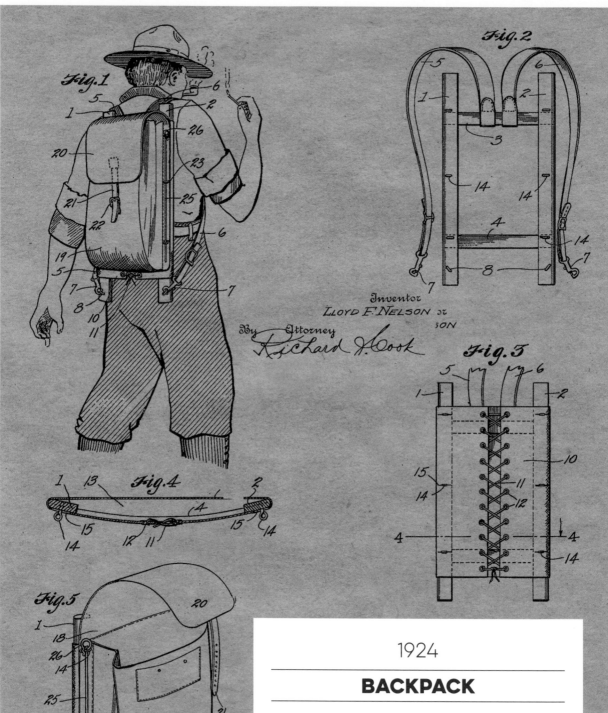

Fig.1

Fig.2

Inventor
LLOYD F. NELSON Jr.
SON

By Attorney
Richard H. Cook

Fig.3

Fig.4

Fig.5

1924

BACKPACK

L. F. NELSON

As is evident in this patent from 1924, the first backpacks were held up by heavy and awkward frames. And yet they were also fragile. In 1936, Victor, Alfred and Gabriel Lafuma lodged the patent for the 'Tyrolian' pack, equipped with a lightweight metal framework. This model was the precursor of the modern backpack.

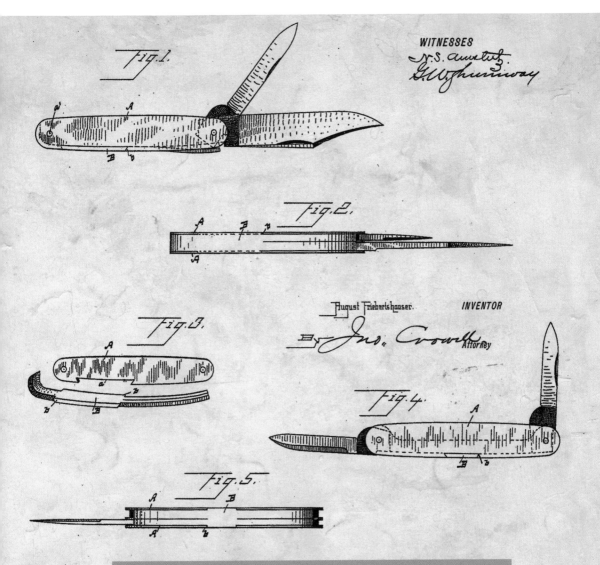

WITNESSES
N. S. Anster.
G. W. Shumway

August Friebertshauser. INVENTOR
By Jnd. Crowell Attorney

Fig.1.

Fig.2.

Fig.3.

Fig.4.

Fig.5.

1886

POCKET KNIFE

A. FRIEBERTSHAUSER

This pocket knife was inspired by the first Swiss Army knives, which date back to 1881. These were first ordered by the Swiss authorities to allow soldiers to open their tinned food and to dismantle their rifles.

The original models included a blade, a bottle opener, a flat screwdriver and a hole punch, all integrated into a handle of black oak.

An exceptional version was made in 2008: it offered 141 functions and 87 tools for a total weight of around 1.4 kg! It holds the record for the most functional Swiss Army knife in the world.

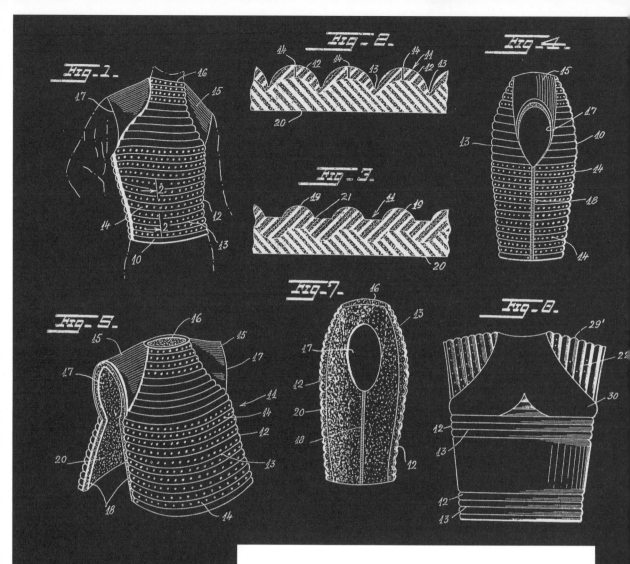

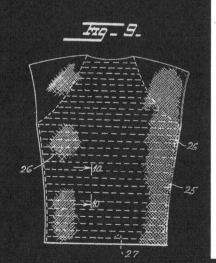

1968

BULLETPROOF VEST

N. J. WATERBURY

The first bulletproof vests were made from silk or linen. Very costly at the time, they were also not very efficient except from a great distance, for minor impacts.

In the 1960s, the US Army created more resistant, lighter and more supple models. These were made from cellular plastic and could be worn in all kinds of situations, as is illustrated by the different sketches on the patent.

The Austrian archduke Franz Ferdinand was wearing a bulletproof vest at the time of his assassination in Sarajevo in 1914. However, the fatal bullet fired by Gavrilo Princip struck him in the neck.

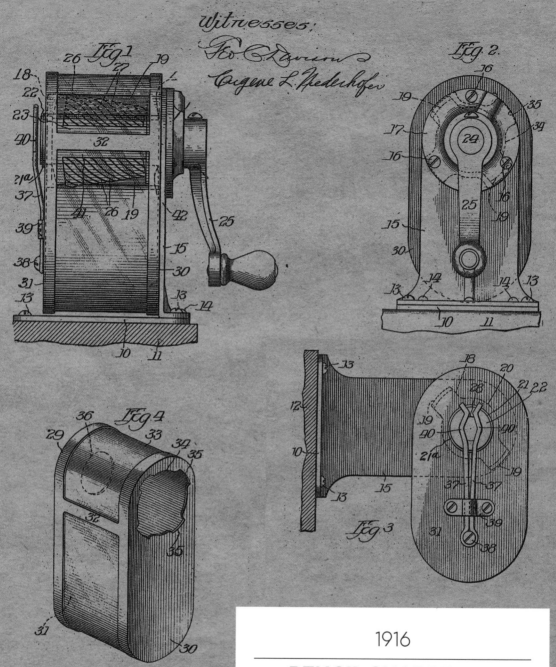

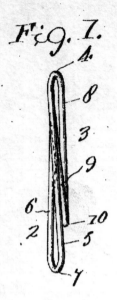

Fig. 1.

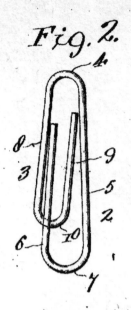

Fig. 2.

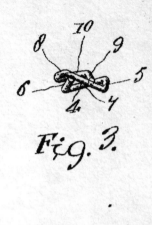

Fig. 3.

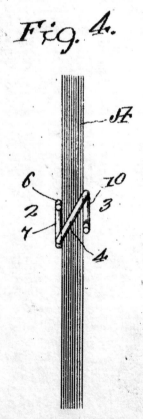

Fig. 4.

1919

PAPERCLIP

B. L. MINK

While the invention of the paperclip dates to the 1890s, the version proposed by Benjamin L. Mink is made from an extendable material, close to that used for springs. It can thus hold more sheets of paper between its rods.

Bizarre fact: in 2006, Kyle MacDonald, a young Canadian, succeeded in trading a simple red paperclip for a house in the town of Kipling, Saskatchewan, thanks to a series of intermediate exchanges based on the successive growth in value of the object offered.

Inventor.
Benjamin L. Mink
by   John S. Barker,
Attorney.

Inventor

ROBERT HELLER

Terry B. Morehouse
Attorney

Fig.1.

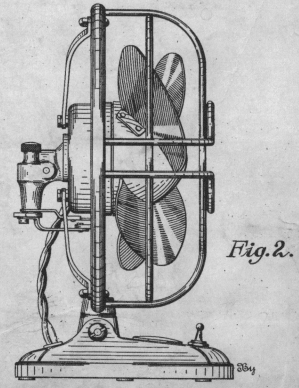

Fig.2.

1936

ELECTRIC FAN

R. HELLER

The patent of Robert Heller concerns a new design for a domestic electric fan, much like those we use today.

Contrary to received wisdom, air conditioning is not such a recent invention. It has existed since 1902, when the American engineer Willis Carrier made the very first air conditioner. Since then, the Carrier Corporation, started by its inventor, has become the global market leader, with more than 4 million units sold every year.

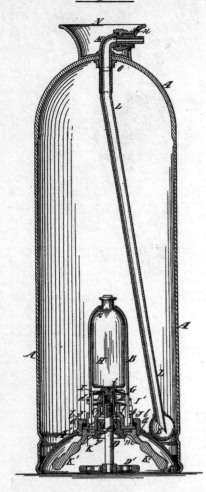

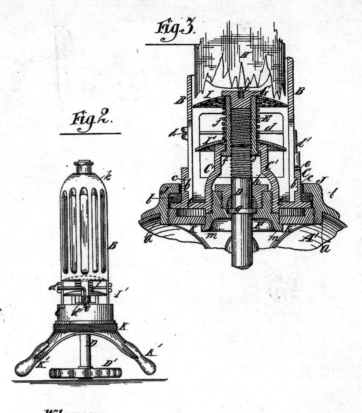

Witnesses :-
Louis M. Hotchkhead,
Fred Haynes

Inventor :-
Almon M. Granger,
by his Attorneys,
Brown & Brown.

1880

FIRE EXTINGUISHER

A. M. GRANGER

The extinguisher of Almon M. Granger is recharged via an opening situated in the bottom of the bottle. Today we still use this system of recharging for CO_2 extinguishers.

In a few years, this type of device will perhaps have changed altogether. In 2015, two students at George Mason University in the United States presented a prototype of a 'sonic' extinguisher. This sends out sound waves at high frequency which 'separate' the oxygen from the flame, thus putting it out.

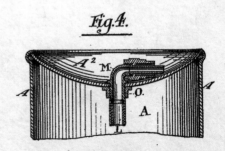

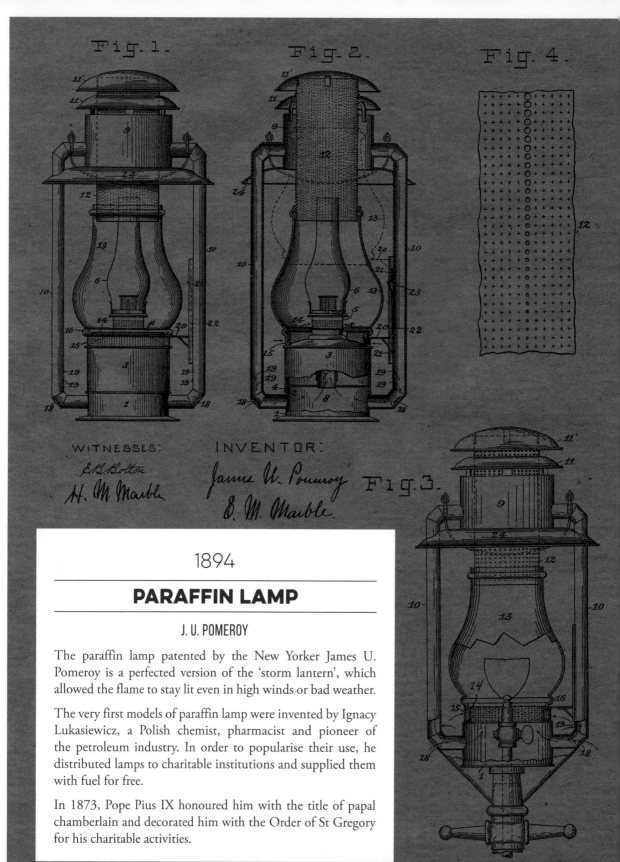

Fig. 1.   Fig. 2.   Fig. 4.

WITNESSES:
E. B. Bolton
H. M. Marble

INVENTOR:
James U. Pomeroy
E. M. Marble

Fig. 3.

1894

PARAFFIN LAMP

J. U. POMEROY

The paraffin lamp patented by the New Yorker James U. Pomeroy is a perfected version of the 'storm lantern', which allowed the flame to stay lit even in high winds or bad weather.

The very first models of paraffin lamp were invented by Ignacy Lukasiewicz, a Polish chemist, pharmacist and pioneer of the petroleum industry. In order to popularise their use, he distributed lamps to charitable institutions and supplied them with fuel for free.

In 1873, Pope Pius IX honoured him with the title of papal chamberlain and decorated him with the Order of St Gregory for his charitable activities.

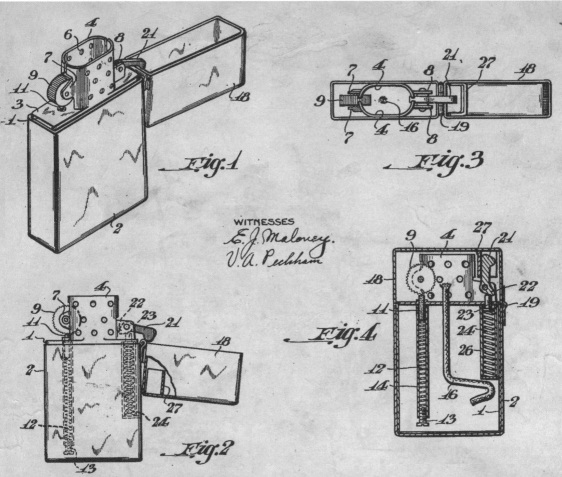

Fig.1

Fig.3

Fig.2

Fig.4

INVENTORS.
George Gimera
George G. Blaisdell
By Brown, Critchlow · Flick
ATTORNEYS

1936

ZIPPO

G. GIMERA

While watching a client of the Bradford Country Club in Pennsylvania light his cigarette with a 'monstrous' lighter in two pieces, George Blaisdell decided to come up with an object that would be reliable, practical and attractive. This would be the Zippo, whose name was inspired by the word 'zipper'.

In 1936, the first patent registered by Blaisdell was signed by a certain George Gimera, who is recognised as the originator of the concept. The design has changed very little since.

Over the past 80 years, the Zippo factory has made more than 500 million pocket lighters, used and collected around the world. Even today, each item is still guaranteed for life and repaired if necessary, whatever its condition.

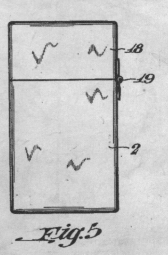

Fig.5

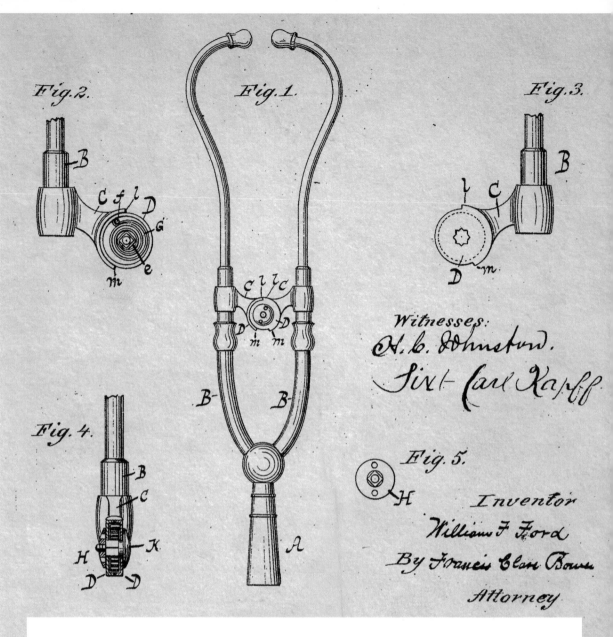

Fig. 2. Fig. 1. Fig. 3.

Witnesses:
H. C. Johnston.
Sixt Carl Kapff

Fig. 4. Fig. 5.

Inventor
William F. Ford
By Francis Clan Bower
Attorney

1882

STETHOSCOPE

W. F. FORD

In 1816, René Laennec, a doctor specialising in respiratory infections, decided out of modesty – or courtesy – to use rolled-up sheets of paper to listen to a patient's chest without putting his head directly on her breast. He found that he could better hear the beating of the heart in this way, and perfected the instrument now known as the stethoscope (from the Greek *stethos*, meaning 'chest').

At that time it was a cylinder made of wood; the later patent of William F. Ford shows a model for two ears, with tubes made of metal.

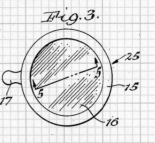

Fig.3.

25
5
15
17
5
16

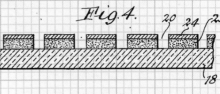

Fig.4.

20 24 23
18

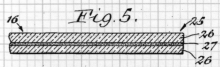

16 *Fig.5.*

25
26
27
26

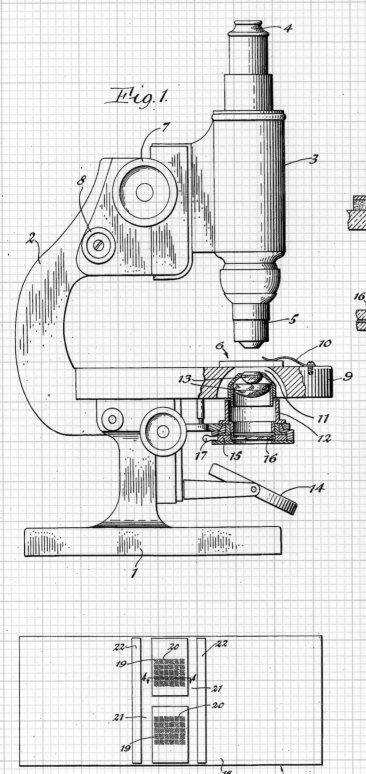

Fig.1.

4
7
3
8
2
5
6
10
13
9
11
12
17 15 16
14
1

Fig.2.

22 20 22
19
4
21
21 20
19
18 6

1948

MICROSCOPE

A. H. BENNETT

It was a Dutch spectacle maker, Zacharias Janssen, who in 1595 had the idea of superimposing two glass lenses within sliding tubes. The magnification thus obtained permitted, for the first time, the observation of elements invisible to the naked eye.

The patent lodged by Alva H. Bennett had the innovation of placing a slanting mirror under the space destined to receive the samples. The user could then vary the intensity of light passing through the lenses.

INVENTOR.
ALVA H. BENNETT
BY *Raymond A. Paquin*
ATTORNEY

81

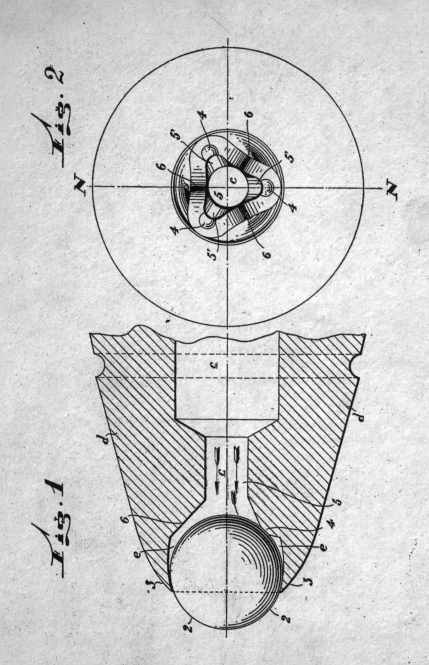

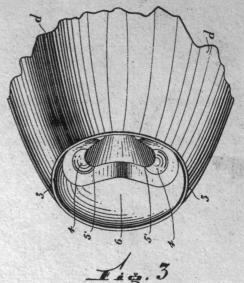

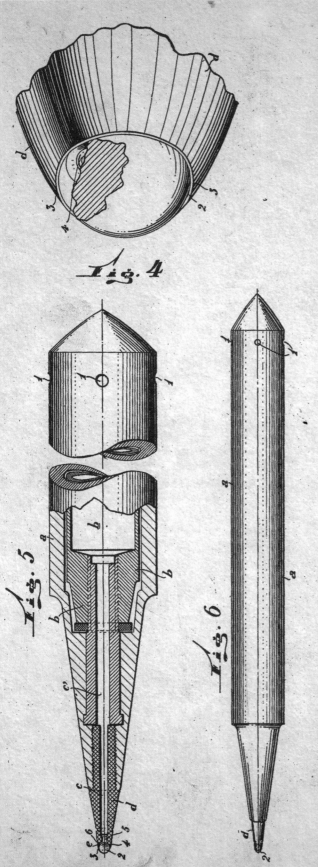

1945

BALLPOINT PEN

L. J. BÍRÓ

As a journalist, the Hungarian László Bíró would get stains on his hands due to the very long time it took for the ink from his pen to dry. He decided therefore to try using printer's ink, which dried faster.

But the ink used in newspaper printing was too thick and sticky to flow easily onto a sheet of paper. Observing the traces left by marbles when children rolled them in a liquid, Bíró had the idea of placing a smaller-sized version of a marble in the head of the pen, to better regulate the flow of ink.

In 1949, Marcel Bich acquired the patent and produced the Bic Cristal, which is still the most frequently bought pen in the world. We call it a Bic or simply a biro.

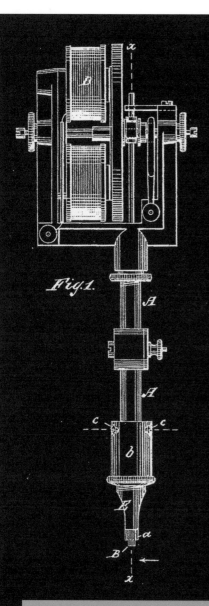

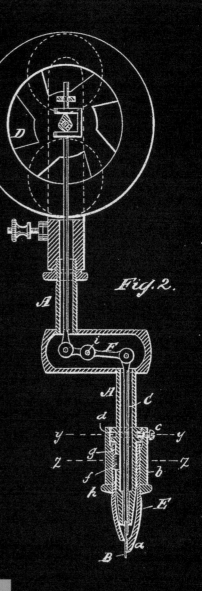

Fig.1.

Fig.3.

Fig.4.

Fig.2.

1891

TATTOO MACHINE

S. F. O'REILLY

Samuel F. O'Reilly was a celebrated New York tattoo artist to whom we owe one of the first models of the electric machine that made modern tattooing possible.

For its invention, he was inspired by the electric pen of Thomas Edison, dating back to 1875. This instrument allowed for the making of several copies of the same document by means of a stencil perforated by a pen driven by an electric motor.

WITNESSES:

Edward Wolff
William Miller

INVENTOR:

Samuel F. O'Reilly

BY

Van Santvoord & Hauff

ATTORNEYS

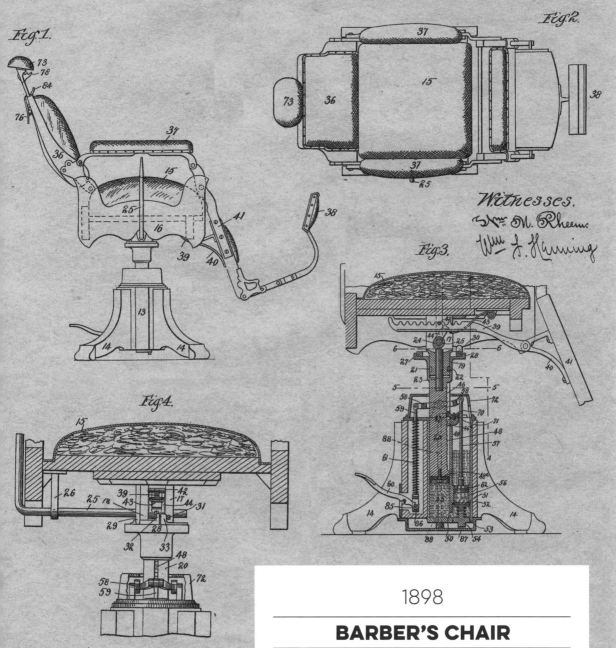

Fig.1.

Fig.2.

Fig.3.

Fig.4.

Witnesses.
Wm. M. Rheem.
Wm. F. Henning

Inventor
Anton J. Rollert,
by Bond Adams Pickard Jackson
Atty's

1898

BARBER'S CHAIR

A. J. ROLLERT

An American of German origin, Ernest Koken was responsible for a notable evolution in the comfort of clients at the barber's. In effect, he invented a reclining armchair then in 1890 he registered the first patent for a model with hydraulic elevation.

The model of Anton J. Rollert allows the activation of the hydraulic system by way of a pedal, whereas the Koken model was activated manually.

Fig. 1.

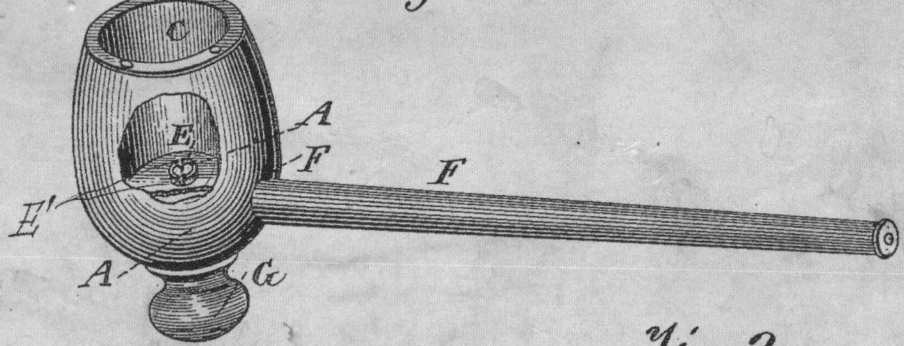

Fig. 2.

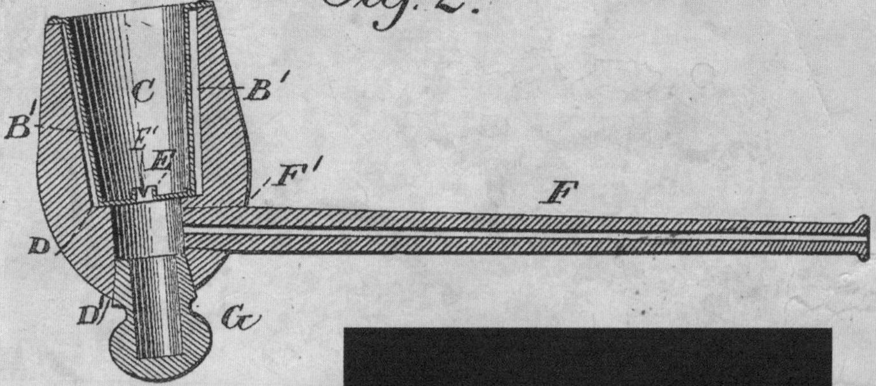

Fig. 3.

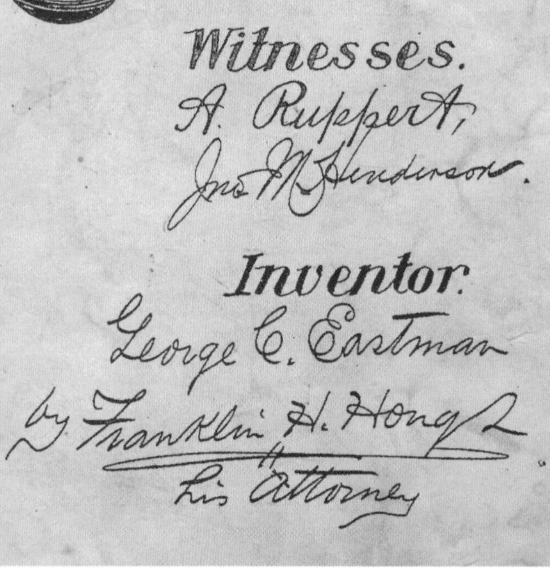

Witnesses.
A. Ruppert,
Jno M Henderson

Inventor.
George C. Eastman
by Franklin H. Hough
his Attorney

1890

PIPE

G. C. EASTMAN

The use of the pipe is an ancient practice in Europe, where models have been found dating back several thousand years. However, it was used not for smoking tobacco but for herbs with therapeutic properties.

Tobacco arrived in Europe thanks to Christopher Columbus, who brought it back from his expedition to America. It was from that moment that pipes began to appear for smoking the new substance.

In 1617, William Baernelts became the first man to manufacture a series of clay pipes. The patent of George C. Eastman shows a model in wood equipped with a removable cup that allowed the user to expel the residue more easily.

SCIENCE AND TECHNOLOGY

Human beings are always searching to surpass the limits imposed upon them: those of time and space, distance and gravity, memory and productivity. The patents which follow are the outcome of innovations with varied objectives: to communicate more easily, to produce more quality and quantity, to capture moments and recall them later, to stay safe and sound in extreme situations, to explore oceans, the skies and even outer space. They all have the aim of allowing us to discover the world around us in different ways.

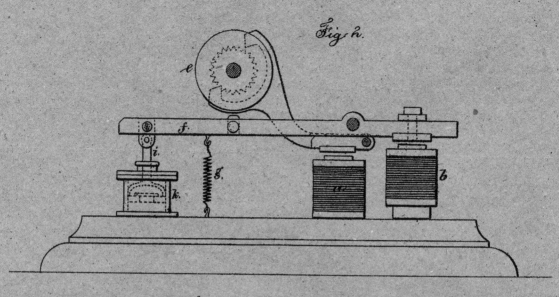

Fig. 2.

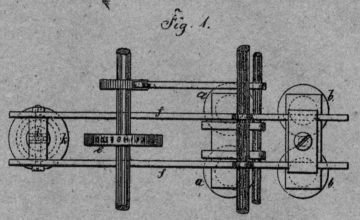

Fig. 1.

Witnesses.
Chos H Smal
Harold Serrell

Inventor
Thos. A. Edison

Lemuel W. Serrell

1872

PRINTING TELEGRAPH

T. A. EDISON

The American Thomas Alva Edison, an insatiable scientist who was nicknamed 'the man of a thousand patents', developed his first inventions while employed as a telegraph operator.

This patent shows a new improvement in the system of telegraphic transmission that Edison had created, and concerns more particularly the printing of telegraphs.

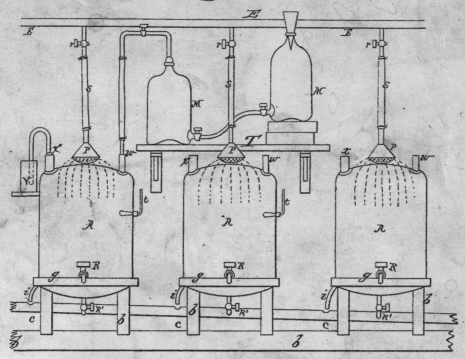

Fig. 1.

Witnesses,

E. Wolf

J. Fielbel

Inventor

Louis Pasteur

By his Attorney

C. M. Keller

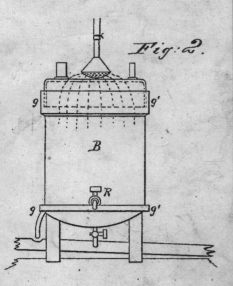

Fig. 2.

1873

INDUSTRIAL BREWING EQUIPMENT

L. PASTEUR

The bombardment of the Museum of Natural History in Paris in 1871 by the German army inspired in Louis Pasteur a lingering bitterness towards Germany, such that he was determined to help French brewers rise to the level of their German counterparts.

He therefore registered a patent for a technique of fermentation that allowed for the making of beer in larger quantities without its quality being affected. The products born of this process should rightly be called 'beers of revenge'!

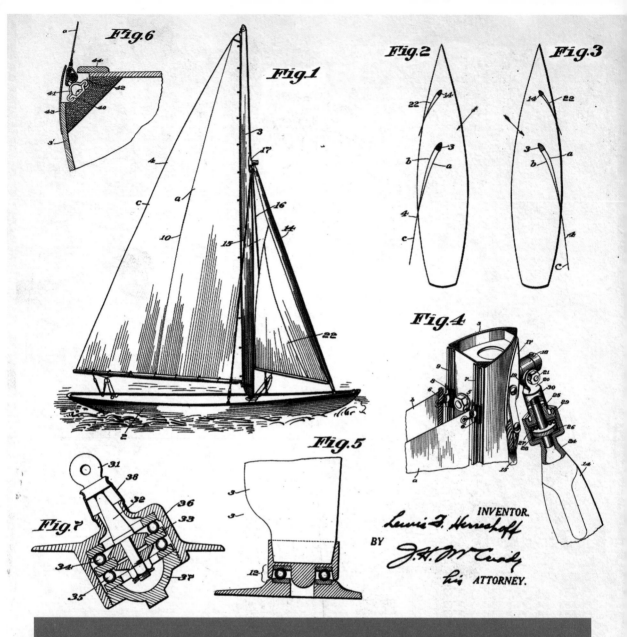

INVENTOR.

Lewis F. Herreshoff

BY

J. H. McCready

his ATTORNEY.

1927

SAILING BOAT

L. F. HERRESHOFF

The existence of sailing boats since ancient times bears witness to humanity's aspiration to travel for the purposes of exploration, commerce, transport, fishing or war. After the industrialisation of the nineteenth century, sailing boats were gradually replaced by steam-powered and then motor-driven versions, the yacht becoming a means of transportation reserved for pleasure cruising and sport.

The patent of the American Lewis Francis Herreshoff concerns racing yachts. The innovations shown are aimed at obtaining better results during regattas.

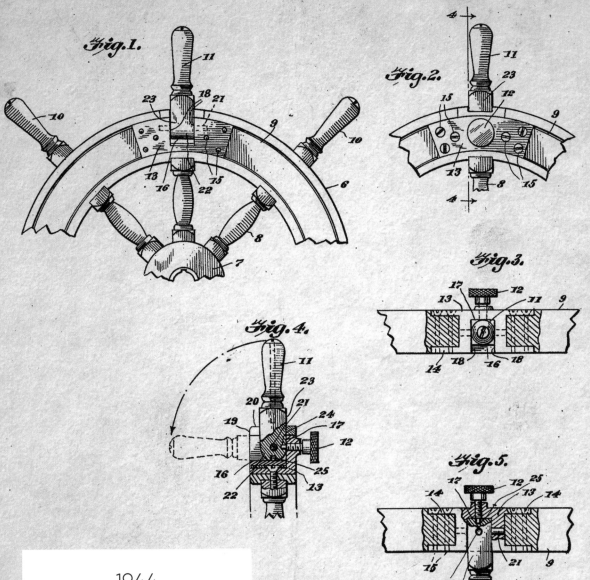

Fig.1.

Fig.2.

Fig.3.

Fig.4.

Fig.5.

1944

SHIP'S WHEEL

A. J. HIGGINS

In the fifteenth century, the first ship's wheels for controlling the rudder made their appearance on caravels and galleons. The version patented by Andrew Jackson Higgins aims to facilitate the rotation of the wheel in both directions, in order to manoeuvre the vessel more precisely and quickly.

Inventor
ANDREW JACKSON HIGGINS

By Mason Fenwick Lawrence
Attorneys

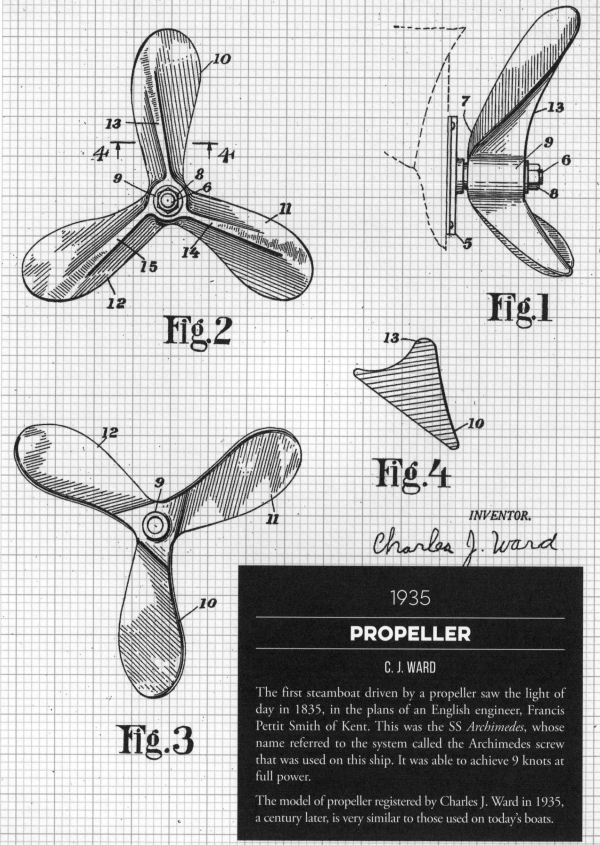

Fig.2

Fig.1

Fig.4

Fig.3

INVENTOR.

Charles J. Ward

1935

PROPELLER

C. J. WARD

The first steamboat driven by a propeller saw the light of day in 1835, in the plans of an English engineer, Francis Pettit Smith of Kent. This was the SS *Archimedes*, whose name referred to the system called the Archimedes screw that was used on this ship. It was able to achieve 9 knots at full power.

The model of propeller registered by Charles J. Ward in 1935, a century later, is very similar to those used on today's boats.

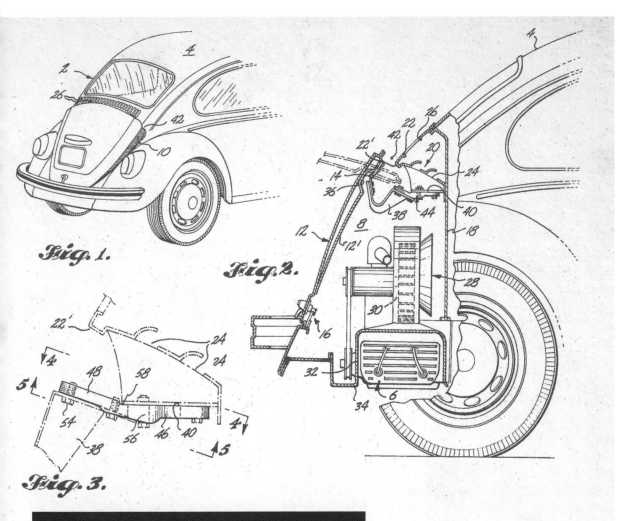

Fig. 1.

Fig. 2.

Fig. 3.

1974

VW BEETLE ENGINE

D. BUSER

Designed by the Austrian engineer Ferdinand Porsche, the Beetle was the first model produced under the German brand Volkswagen ('the people's car').

Its phenomenal success is partly explained by vast media coverage which made the little car into the star of six films produced by Disney between 1968 and 2005. The car was called Herbie, and its first comedic outing was in *The Love Bug*.

The patent lodged by Dans Buser is for the addition of ventilation slots in the rear section of the vehicle, in order to prevent the overheating of the engine.

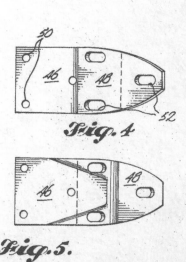

Fig. 4.

Fig. 5.

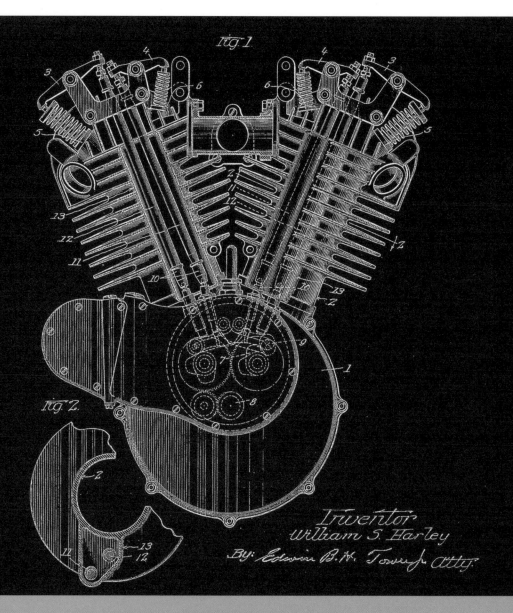

Fig. 1.

Fig. 2.

Inventor
William S. Harley
By: Edwin B. H. Tower Jr. Atty.

1923

MOTORCYCLE ENGINE

W. S. HARLEY

The first patent for a 'high-speed bicycle' is attributed to a Parisian engineer, Louis-Guillaume Perreaux, who in 1868 installed a steam engine in a wooden body.

The patent shown here was lodged by William Harley, founder of Harley-Davidson in 1903. The brand was quick to gain renown, supplying the US Army with two-wheelers during World War One. Today the company is still one of the most prestigious makers of motorcycles in the world.

This engine contributed to the brand's success: lighter and more compact than its predecessors, it offered better transmission while avoiding problems of overheating.

Inventors:
William S Harley
Arthur R Constantine,
By Wilkinson, Huxley, Byron+Knight
Attys

Fig. 1

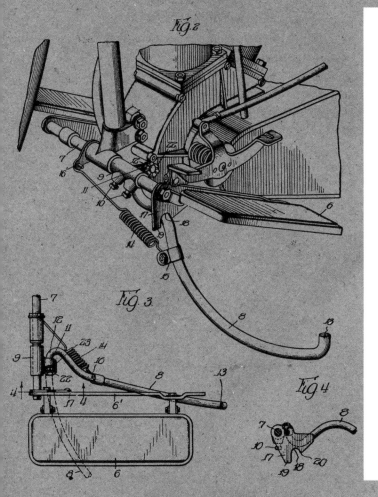

Fig. 2

Fig. 3

Fig. 4

1928

MOTORCYCLE STAND

W. S. HARLEY

This is one of the first motorcycle stands which could be operated with a single kick, which made it possible to easily keep the machine upright when stationary.

In a constant effort to improve user comfort, Harley-Davidson regularly lodges new patents relating to maintaining the seat, quietening the muffler and so forth.

Unfortunately, the financial crash of 1929 had a serious effect on the company, which had to let go half its employees and reduce its production by the same amount. It was not until the 1940s that the brand recovered and began its regrowth.

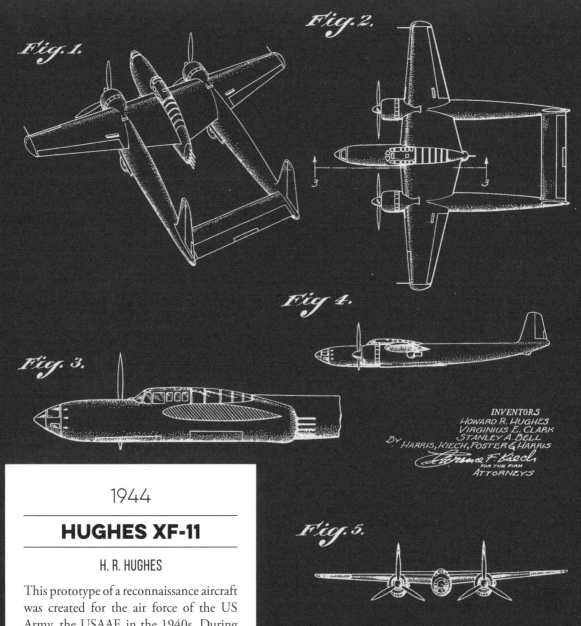

Fig. 1.

Fig. 2.

Fig. 4.

Fig. 3.

INVENTORS
HOWARD R. HUGHES
VIRGINIUS E. CLARK
STANLEY A. BELL
BY HARRIS, KIECH, FOSTER & HARRIS
Clarence F. Kiech
FOR THE FIRM
ATTORNEYS

Fig. 5.

Fig. 6.

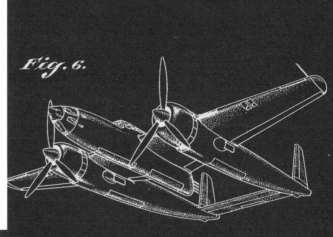

1944

HUGHES XF-11

H. R. HUGHES

This prototype of a reconnaissance aircraft was created for the air force of the US Army, the USAAF, in the 1940s. During the first flight, on 7 July 1946, the aircraft crashed, seriously injuring the pilot, who was also the holder of the patent.

This was none other than Howard R. Hughes, one of the richest men in the USA at the time, especially famous as a producer of big-budget films at the end of the 1920s. Passionate about aviation, he established several world speed records and had other models built, such as the Hughes H-1 Racer and the H-4 Hercules.

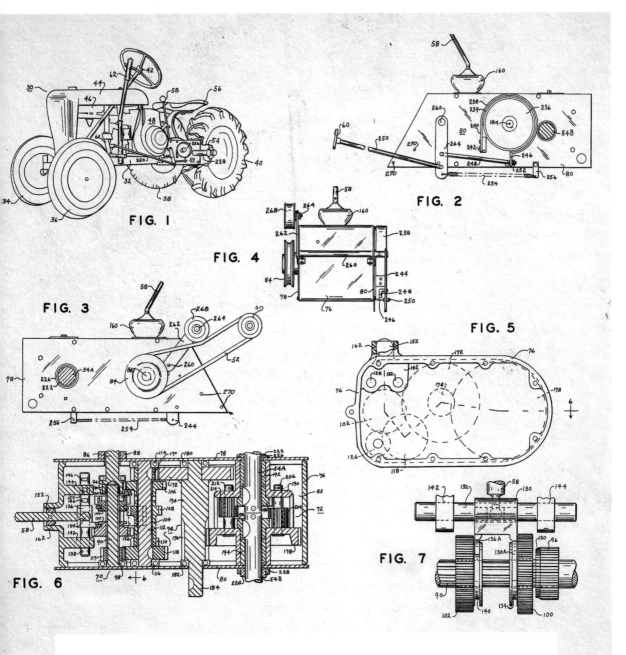

1961

TRACTOR

C. E. POND

Fyodor Blinov, a Russian inventor, developed history's first tracked vehicle in 1877. Concerned about helping farmers, he adapted his system for the steam tractors of the era, with his continuous-track farm tractor being publicly unveiled at an exhibition in 1896.

The patent of Cecil E. Pond shows a new mechanism of electric transmission intended for smaller models, ideal for small farms or private users.

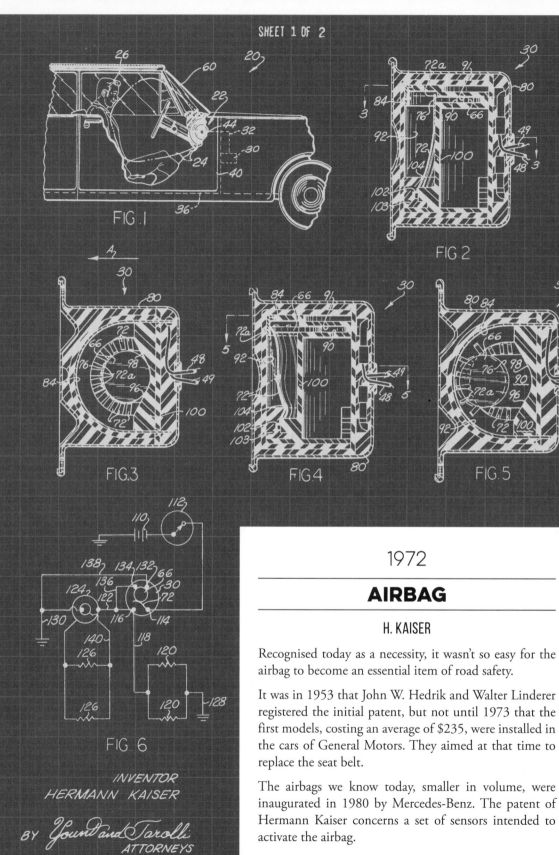

FIG.1

FIG.2

FIG.3

FIG.4

FIG.5

FIG.6

INVENTOR
HERMANN KAISER

BY *Yound and Tarolli*
ATTORNEYS

1972

AIRBAG

H. KAISER

Recognised today as a necessity, it wasn't so easy for the airbag to become an essential item of road safety.

It was in 1953 that John W. Hedrik and Walter Linderer registered the initial patent, but not until 1973 that the first models, costing an average of $235, were installed in the cars of General Motors. They aimed at that time to replace the seat belt.

The airbags we know today, smaller in volume, were inaugurated in 1980 by Mercedes-Benz. The patent of Hermann Kaiser concerns a set of sensors intended to activate the airbag.

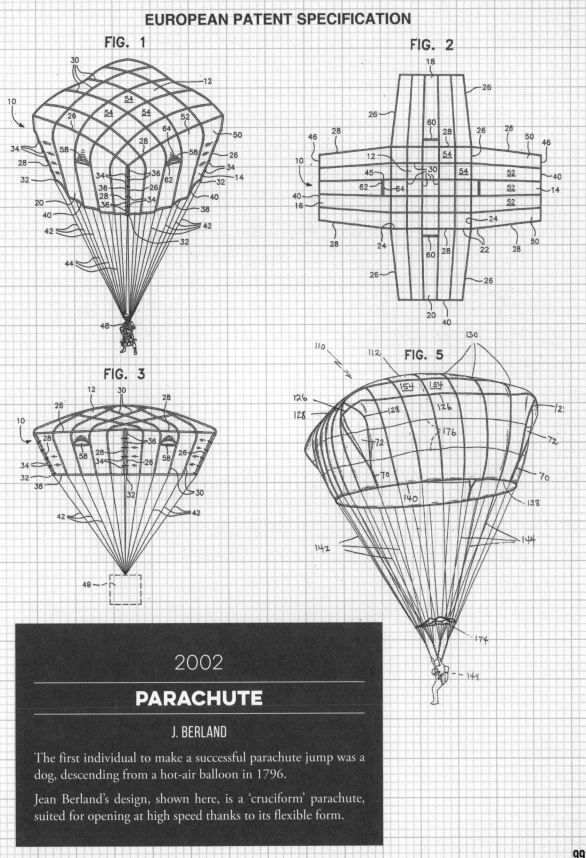

FIG. 1

FIG. 2

FIG. 3

FIG. 5

2002

PARACHUTE

J. BERLAND

The first individual to make a successful parachute jump was a dog, descending from a hot-air balloon in 1796.

Jean Berland's design, shown here, is a 'cruciform' parachute, suited for opening at high speed thanks to its flexible form.

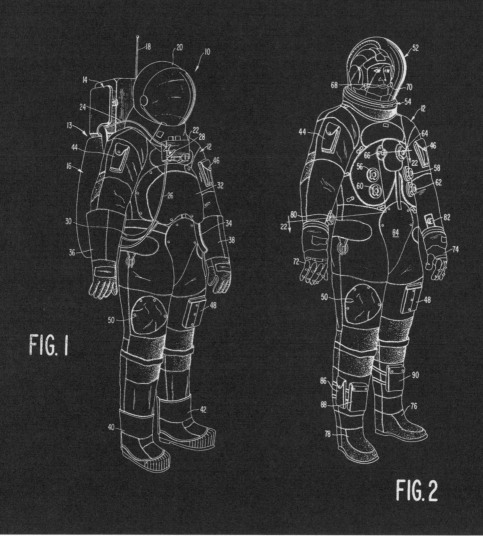

FIG. 1

FIG. 2

1973

A7L SPACE SUIT

L. F. SHEPARD

This model of space suit was designed for the missions of the Apollo programme. Although the suits were made to measure and used from 1968, the patent was not registered until 1973. The suits weighed 72 kilos, but with lunar gravity being six times weaker than that on Earth, the astronauts had the impression of wearing 'just' 14 kilos of kit.

The upper part of the suit includes a helmet equipped with a sun shield and a tinted visor to protect the eyes from solar and ultraviolet radiation.

The reserves of oxygen, electricity and cooling water allowed for a maximum of 6.5 hours of autonomous movement.

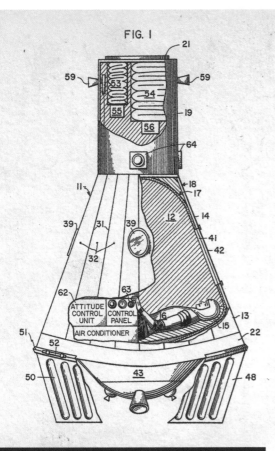

FIG. 1

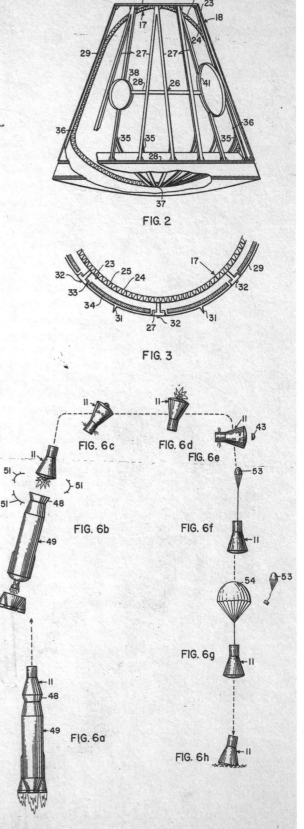

FIG. 2

FIG. 3

FIG. 6c

FIG. 6d

FIG. 6e

FIG. 6b

FIG. 6f

FIG. 6g

FIG. 6a

FIG. 6h

1963

SPACE CAPSULE

M. A. FAGET

This patent was registered on behalf of NASA by the American Maxime A. Faget and five other scientists. It shows one of the first space capsules capable of being launched into orbit before returning to Earth.

It was with this type of model that Yuri Gagarin achieved the first human flight in space in 1961. Since then, more than 500 people, over 50 of them women, have been sent into space. Twelve of them have walked on the moon.

INVENTORS
M.A. FAGET   W.S. BLANCHARD, JR.
A.J. MEYER, JR. A.B. KEHLET
R.G. CHILTON   J.B. HAMMACK
C.C. JOHNSON, JR.

BY

ATTORNEYS

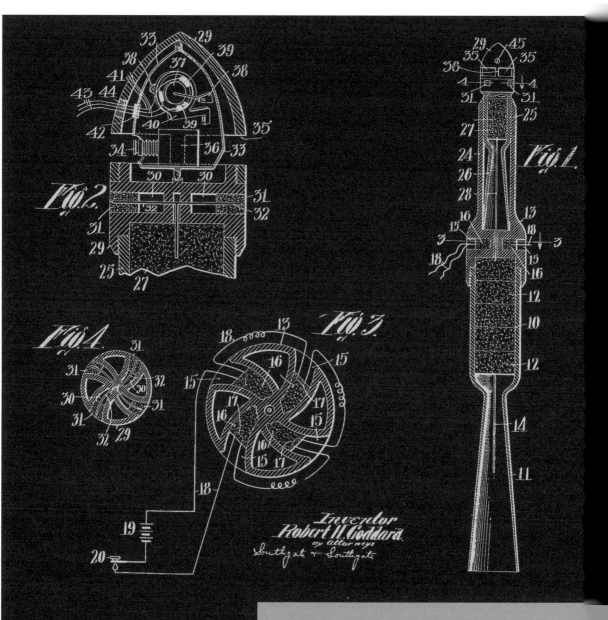

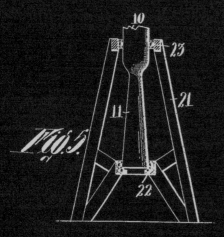

1914

ROCKET

R. H. GODDARD

It was after reading *The War of the Worlds*, the science-fiction novel by H. G. Wells, that the American Robert H. Goddard, engineer and physician, dedicated himself to the 'conquest' of space.

He developed the first powered rocket driven by liquid propellant, though at the first attempt the rocket attained a height of just 12.5 metres!

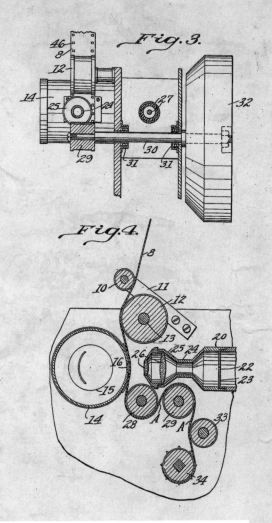

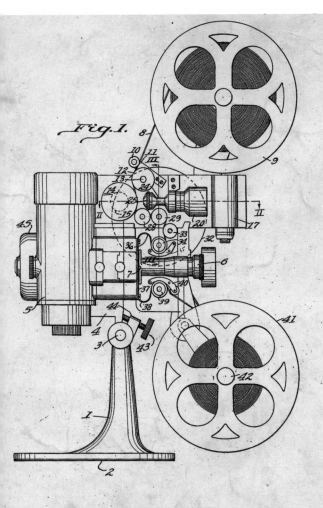

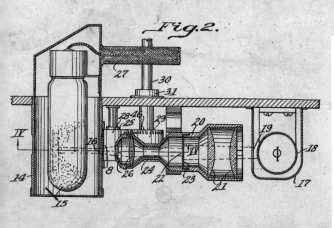

1936

FILM PROJECTOR WITH SOUND

T. LINDENBERG JR

The Jazz Singer is considered the first feature-length 'talkie' in film history, with 281 words spoken by the actors. Released on 6 October 1927, it contributed greatly to the success of this new form of cinema.

In 1933, Theodore Lindenberg Jr invented this model of cinema projector. The device can also be used by individuals and in small venues.

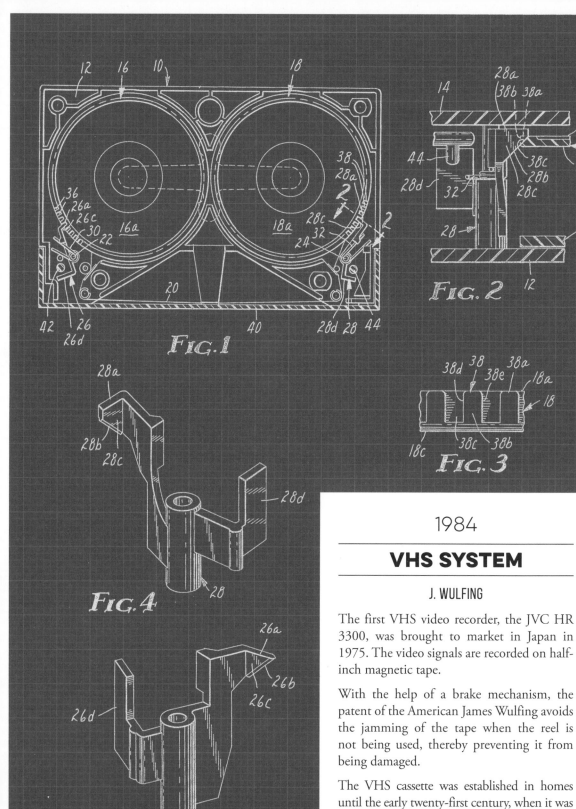

FIG. 1

FIG. 2

FIG. 3

FIG. 4

FIG. 5

1984

VHS SYSTEM

J. WULFING

The first VHS video recorder, the JVC HR 3300, was brought to market in Japan in 1975. The video signals are recorded on half-inch magnetic tape.

With the help of a brake mechanism, the patent of the American James Wulfing avoids the jamming of the tape when the reel is not being used, thereby preventing it from being damaged.

The VHS cassette was established in homes until the early twenty-first century, when it was superseded by the DVD. Sales of DVD players surpassed those of video players in 2003.

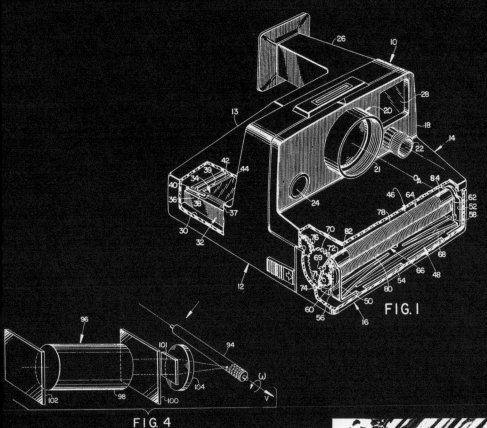

FIG. I

FIG. 4

FIG. 3

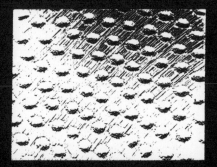

FIG. 2

1980

POLAROID LAND CAMERA 1000

H. S. FRIEDMAN

The Polaroid company was created by Edwin H. Land, an American scientist who invented a polarising material used particularly in the field of optics.

The business is best known for bringing to market cameras with instantaneous development. Owing to competition from digital photography, Polaroid ceased production of its cameras in 2007, before launching new models in 2011, following the resurgence in film photography.

The model shown on this patent is that used by Andy Warhol to take his famous shots of friends and other celebrities. In 2013, one of the artist's cameras was put on sale on eBay for the sum of $50,000.

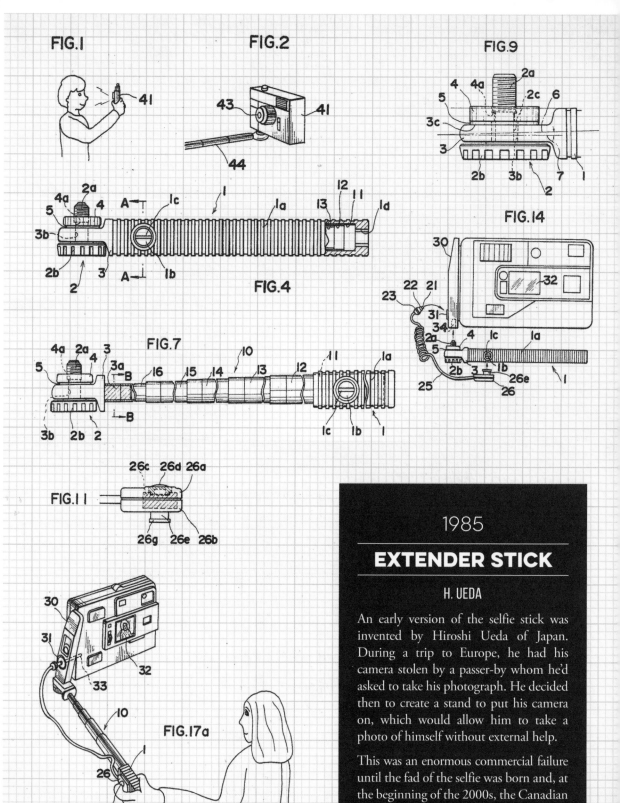

FIG.1
FIG.2
FIG.9
FIG.4
FIG.7
FIG.11
FIG.14
FIG.17a

1985

EXTENDER STICK

H. UEDA

An early version of the selfie stick was invented by Hiroshi Ueda of Japan. During a trip to Europe, he had his camera stolen by a passer-by whom he'd asked to take his photograph. He decided then to create a stand to put his camera on, which would allow him to take a photo of himself without external help.

This was an enormous commercial failure until the fad of the selfie was born and, at the beginning of the 2000s, the Canadian Wayne Fromm picked up the idea to adapt it for the era of mobile phones.

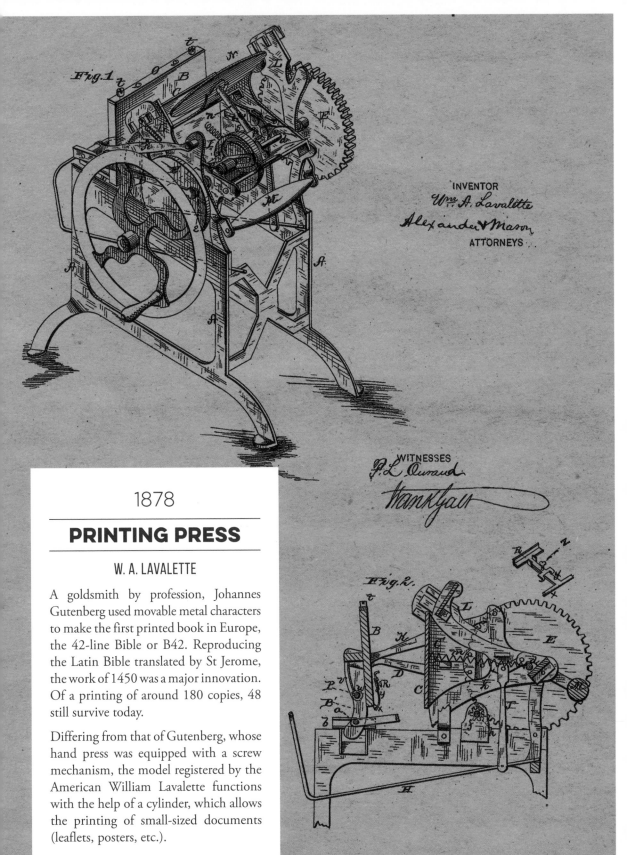

INVENTOR
W^m. A. Lavalette
Alexander & Mason
ATTORNEYS

WITNESSES
F. L. Durand
Frank Gall

Fig. 1

Fig. 2

1878

PRINTING PRESS

W. A. LAVALETTE

A goldsmith by profession, Johannes Gutenberg used movable metal characters to make the first printed book in Europe, the 42-line Bible or B42. Reproducing the Latin Bible translated by St Jerome, the work of 1450 was a major innovation. Of a printing of around 180 copies, 48 still survive today.

Differing from that of Gutenberg, whose hand press was equipped with a screw mechanism, the model registered by the American William Lavalette functions with the help of a cylinder, which allows the printing of small-sized documents (leaflets, posters, etc.).

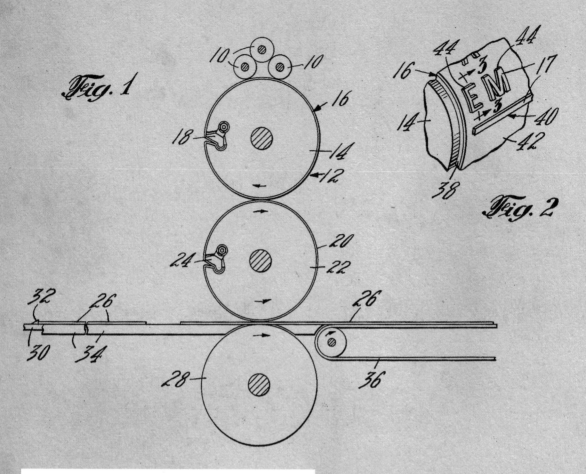

Fig. 1

Fig. 2

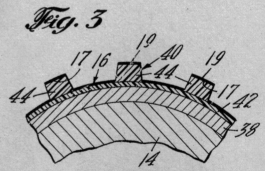

Fig. 3

INVENTOR.
HARRIS JAY WOLBERT
BY *Frank R Presta*
George W. Reiber
ATTORNEYS

1965

OFFSET PRINTING

H. J. WOLBERT

The quality obtained and the low costs of offset production made it a major method of printing in the second half of the twentieth century. It replaced the movable lead type used since 1450. The process offers great flexibility, with print runs varying between 1,000 and several hundred thousand copies.

While offset printing is traditionally achieved by the repulsion of two opposing substances – water and ink – the patent of American Harris Jay Wolbert puts forward a process that does not resort to water: the unprinted zones are covered in silicon in order to prevent ink from being deposited there.

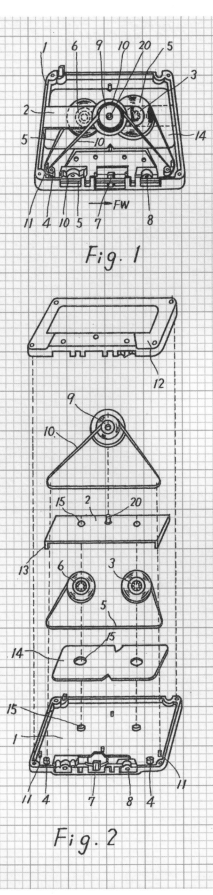

Fig. 1

Fig. 2

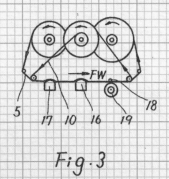

Fig. 3

INVENTOR.
KIYOSHI HATA

BY *Eugene Geoffroy Jr*

ATTORNEY

1972

AUDIO CASSETTE

K. HATA

The audio cassette (also called the 'minicassette') was offered to the public at the same time as the first tape recorder (the E1-3-300) in 1963.

The product met with huge success, then sales declined in the 1980s due to the birth of the CD. The latter offered a stunning sound quality, perfectly reproducing the original piece.

However, certain artists continue to use this format, such as the French group Daft Punk, who brought out a limited edition series of cassettes for their last album in 2013.

The cassette design by Kiyoshi Hata has two tapes. This type was mostly used for telephone answering machines. The first tape allowed the voice message to issue a greeting, while the second recorded the received messages.

INVENTORS
TOKUJI NEGISHI AND MOTOMASA YOSHIDA
BY
Linton and Linton
ATTORNEYS

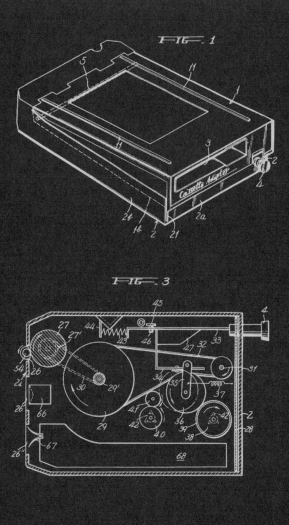

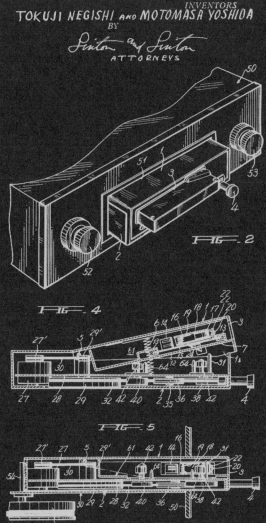

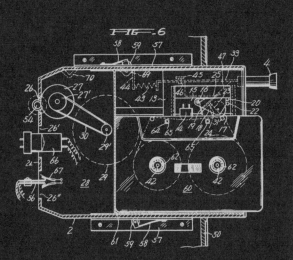

1975

CAR RADIO

T. NEGISHI AND M. YOSHIDA

In Europe, the first radio receiver able to be connected to a car battery, the Blaupunkt Autosuper AS5, went on sale in 1932. It was a simple cube-shaped box, fixed to the steering column. The only two adjustments available were the volume and the choice of station.

This 1975 model, designed by Tokuji Negishi and Motomasa Yoshida, can play or record music on an audio cassette.

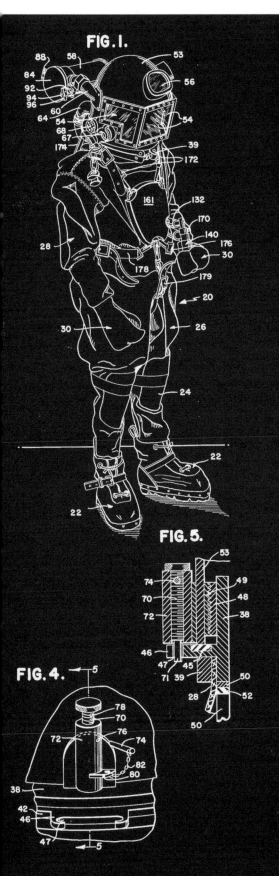

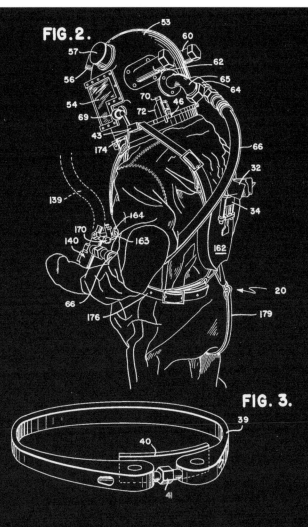

1953

SCUBA DIVING SUIT

E. D. BUIE

The scuba diving suit was originally developed to help miners in the event of a gas explosion or sudden surge of groundwater. The first examples, created in 1860, allowed the user to stay underwater for 30 minutes at a depth of up to 10 metres.

The scuba suit of Emerson D. Buie is an improvement on previous versions, because his model was much lighter. Intended for the US Army, it was principally used for missions to eliminate underwater mines and for demolition work.

EVERYDAY OBJECTS

The objects we use every day are the fruit of technological innovations which have shaken up our lives and our societies. The alarm clock, the bottle opener, the umbrella, the fork, the hairdryer, the light bulb, the computer – they seem to have always been with us, to the point that we rarely even stop to ask ourselves what we'd do without them. The following pages are an opportunity to explore the history of their invention and evolution, and perhaps to see them in a new light.

Fig.1.

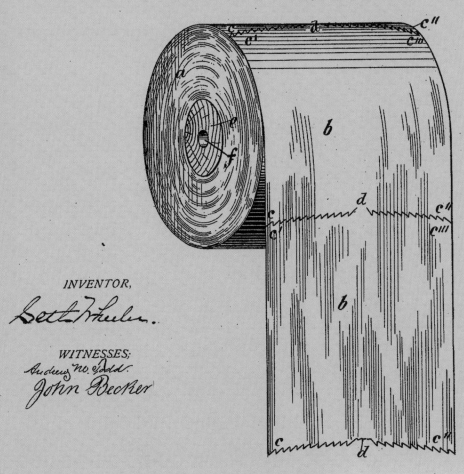

INVENTOR,

Seth Wheeler.

WITNESSES;

Andrew No. Todd.
John Becker

Fig. 2.

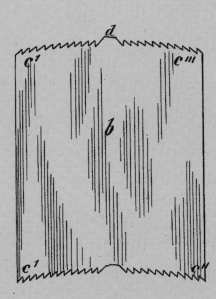

1891

TOILET PAPER ROLL

S. WHEELER

This patent is one of the first to have been registered for the toilet paper roll, invented a decade earlier and developed by the American society APWP Co.

This type quickly became the norm, and 22 billion rolls of toilet paper are now sold every year around the world. Each of them has an average of 140 to 160 sheets.

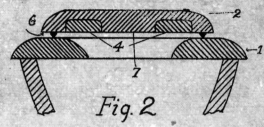

Fig. 2

Inventor:
Herbert Cumming MacDonald

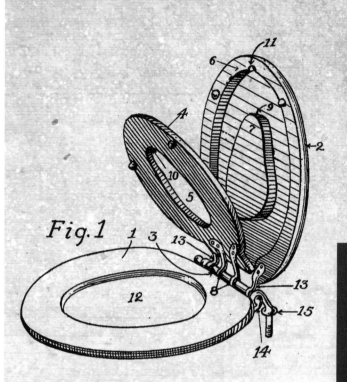

Fig. 1

1936

TOILET SEAT AND LID

H. C. MacDONALD

While this invention dates back to the eighteenth century, MacDonald's patent shows the toilet seat familiar to us today. However, it does have a remarkable innovation: two seating options.

The little one in MacDonald's design can be installed for children and for people of smaller dimensions, though today there are plastic 'reducer rings' available which serve the same purpose.

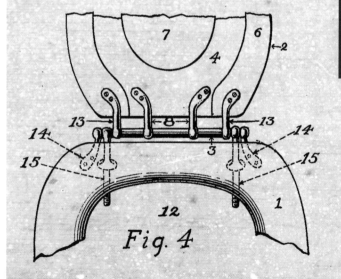

Fig. 4

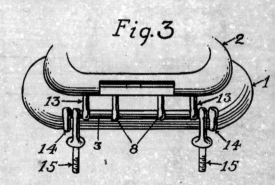

Fig. 3

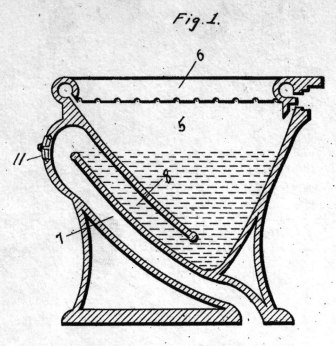

Fig. 1.

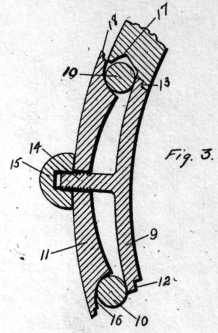

Fig. 3.

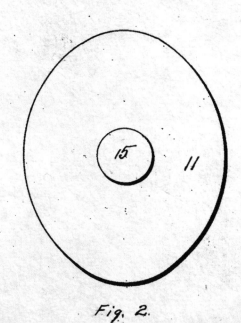

Fig. 2.

Inventor
William E. Gibson
by Geo. E. Tew.
Attorney

1909

TOILET SYPHON

W. E. GIBSON

This invention considerably improved the working of toilets. Evacuation is facilitated by a space at the level of the syphon.

The invention of the flush itself is in fact thanks to an English poet, John Harington, who devised the modern system of flushing in 1595 for his godmother, Queen Elizabeth I. The odours that emanated from her existing facility were so unpleasant that the sovereign had asked him to find a remedy. What he came up with was a rudimentary system using a water tank placed on the roof.

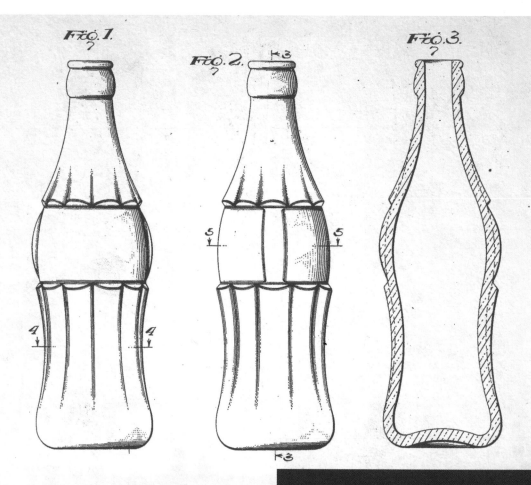

FIG. 1.   FIG. 2.   FIG. 3.

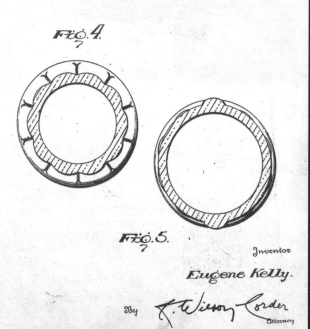

FIG. 4.

FIG. 5.

Inventor

Eugene Kelly.

By K. Wilson Corder

Attorney

1937

COCA-COLA BOTTLE

E. KELLY

The model of the 'contour bottle' was created in 1915 by the Root Glass Company on behalf of Coca-Cola. This patent attests to the transfer of ownership between the two companies.

The specifications imposed precise criteria: the bottle had to be recognisable by touch – in the dark or even lying broken on the ground – and had to be gripped easily.

The bottle kept its iconic shape, but today its use has declined with the introduction of different sizes and materials.

More than 1.9 billion servings of Coca-Cola products are sold every day.

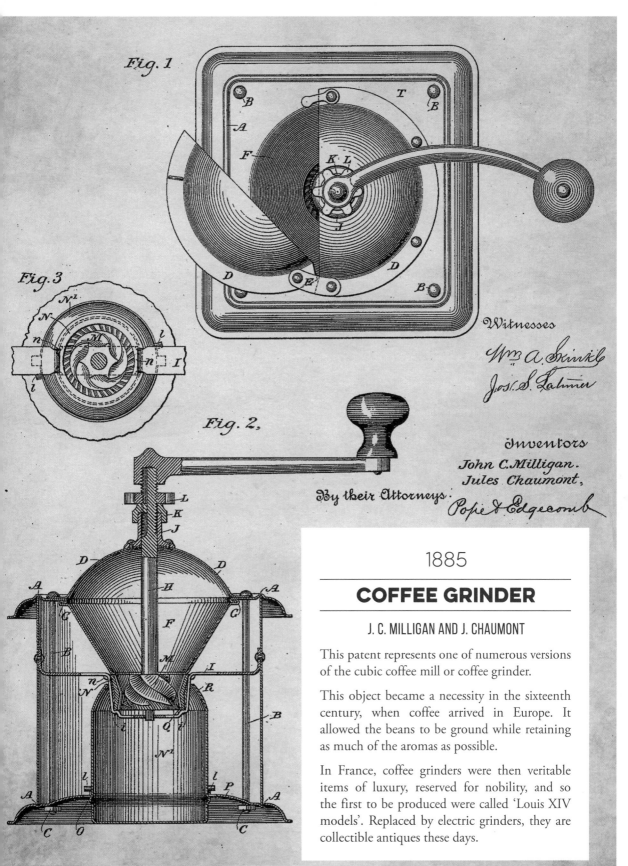

Fig. 1

Fig. 3

Fig. 2,

1885

COFFEE GRINDER

J. C. MILLIGAN AND J. CHAUMONT

This patent represents one of numerous versions of the cubic coffee mill or coffee grinder.

This object became a necessity in the sixteenth century, when coffee arrived in Europe. It allowed the beans to be ground while retaining as much of the aromas as possible.

In France, coffee grinders were then veritable items of luxury, reserved for nobility, and so the first to be produced were called 'Louis XIV models'. Replaced by electric grinders, they are collectible antiques these days.

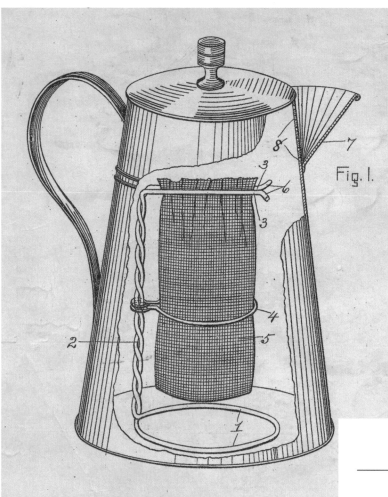

Fig. 1.

WITNESSES:

J. Donsbach.
E. Bakeman.

INVENTOR
James H. Myers
BY
Mosher & Curtis
ATTORNEYS

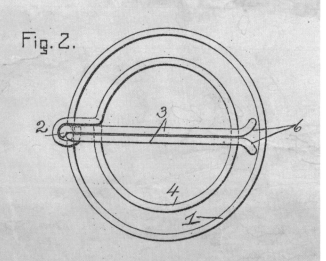

Fig. 2.

1910

PERCOLATOR

J. H. MYERS

The percolator for brewing coffee had been invented by the American physicist and soldier Benjamin Thompson, Count Rumford, in 1806.

The Myers version shown here is more sophisticated as its clamps stop the coffee bag from falling or opening.

Did you know that the coffee craze in the West in the sixteenth century was far from being uncontroversial? Some Catholics even demanded that Pope Clement VIII should prohibit the 'dark drink of Islam'. But tasting the beverage himself, he declared: 'The aroma of coffee is far too agreeable to be the work of the Devil… It would be a pity for the Muslims to have it all to themselves.'

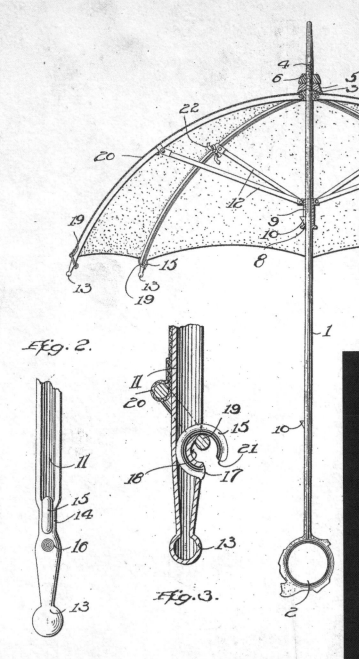

Fig. 1.

4
6
5
3
22
11
22
20
12
12
19
9
10
15
8
13
13
19
13
19
15
13
1
15
13

Fig. 2.

11
20
11
15
15
14
16
19
21
18
17
10
13
13
13
2

Fig. 3.

WITNESSES:

Jr. Niees

Jw. L. McCathran.

INVENTORS
James Ahern
Lester Frederick

By E.E. Trooman, Attorney

1912

UMBRELLA

J. AHERN AND L. FREDERICK

This patent proposes various innovations relating to opening the umbrella and fixing the material on the struts.

The umbrella had been invented in China several centuries earlier. It arrived in Europe in the eighteenth century apparently owing to the English traveller Jonas Hanway, who proudly showed off the object in the streets of London, in spite of the ridicule of passers-by.

The umbrella was then subject to numerous innovations. Jean Marius, a Parisian craftsman, invented a folding version, for example.

At the time, coachmen saw it as unfair competition, while among high society its use was seen as implying that one couldn't afford a car.

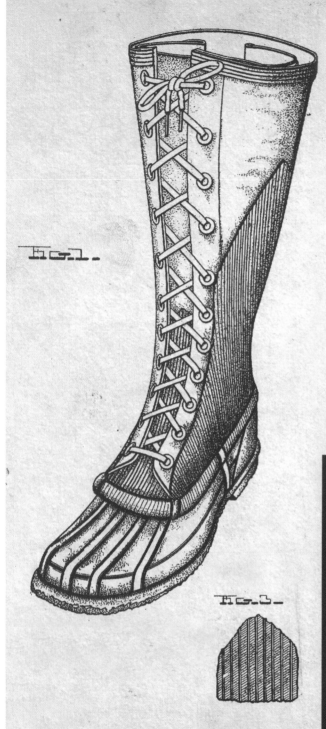

FIG. 1.

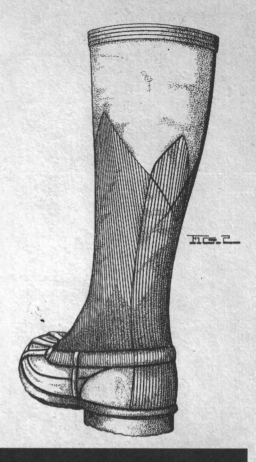

FIG. 2.

FIG. 3.

1926

RUBBER BOOTS

E. S. BOTT

This patent shows one of the first types of rubber boots, to be used for fishing or hunting.

From the Renaissance up until the French Revolution, it was thought that only men should wear thigh-length boots. The boots made fashionable among British gentlemen by the Duke of Wellington in the early 1800s originally stopped mid-calf and were made of leather.

It was Hiram Hutchinson, an American of British origin, who in 1853, after having bought Charles Goodyear's patents of vulcanisation, adapted the flexible rubber for boots.

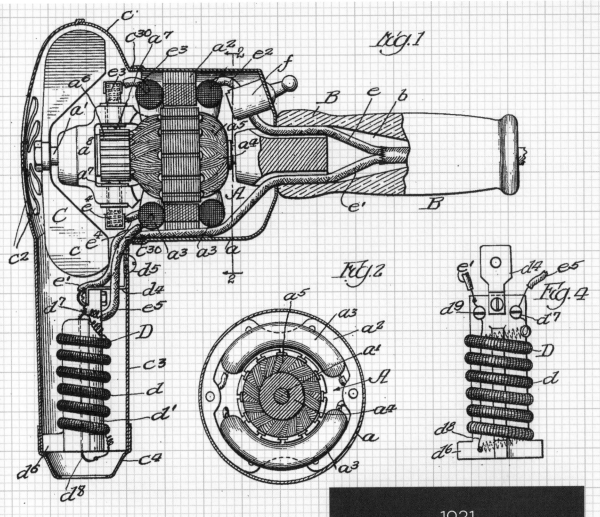

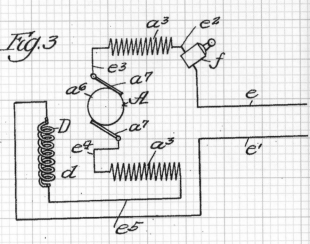

1931

HAIRDRYER

E. NIELSEN

This 1931 patent concerns one of numerous models registered after the invention of the manual hairdryer in 1926 by an engineer for Calor named Léo Trouilhet. His prototype was then called the 'electric shower of hot and cold air'. It improved the mechanical system developed by Alexandre Godefroy in 1886, which conveyed hot air produced by a gas cooker to a bonnet placed over the hair.

It was Jean Mantelet, another engineer, this time from Moulinex, who modified the hairdryer's shape, inspired by that of the electric drill.

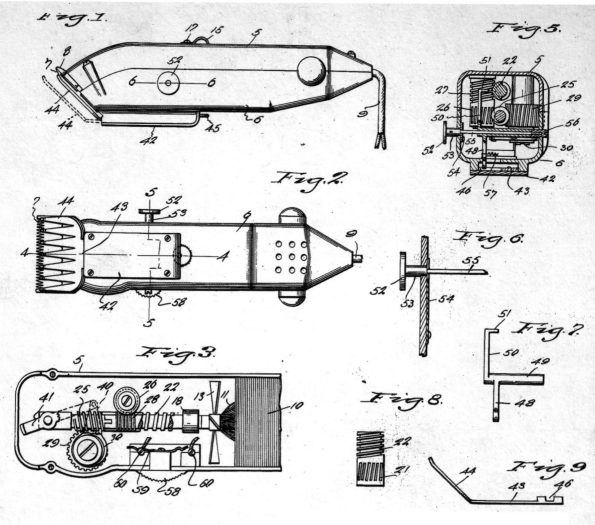

Fig.1.

Fig.5.

Fig.2.

Fig.6.

Fig.7.

Fig.8.

Fig.3.

Fig.9.

Fig.4.

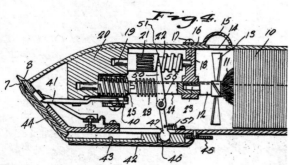

Inventors

Ralph Greco
George Getz
Frank Marino

By *Clarence A. O'Brien*

Attorney

1935

BARBER'S CLIPPERS

R. GRECO, G. GETZ AND F. MARINO

On 14 October 1919, the American Leo Wahl registered the patent for the first clippers with an integrated electromagnetic motor. Less than a decade later, thousands of such clippers had a place in the salons of barbers and hairdressers in the USA.

The model shown here is not much different from those developed by the Leo Wahl company. But the latter would be alone in mastering the manufacture of portable (and nearly silent) devices on an industrial scale.

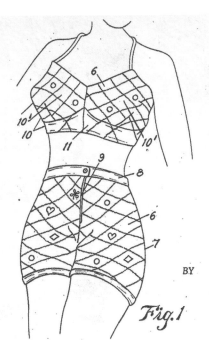

Fig.1

INVENTOR,
Margaret R. Cobb

BY

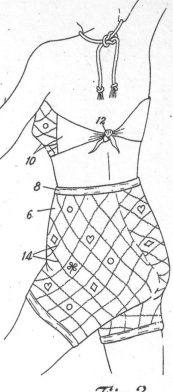

Fig.2

Fig.3

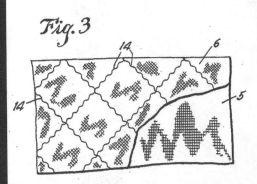

1944

BATHING SUIT

M. R. COBB

Before the arrival of the bikini, modesty required the wearing of the two-piece bathing suit. It was mainly made of jersey material, like the model presented here, and thus it did not lose its shape after bathing.

It was Micheline Bernardini, celebrated nude dancer at the Casino de Paris, who in 1946 modelled the first bikini at the Molitor swimming pool. No other model of the era agreed to be seen dressed in this bathing suit, 'smaller than the smallest bathing suit in the world'.

The word 'bikini' comes from the Bikini Atoll, where a nuclear explosion had taken place a few days before.

Fig.4

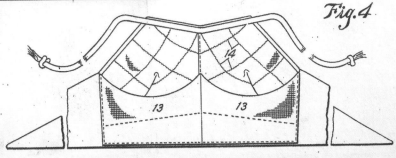

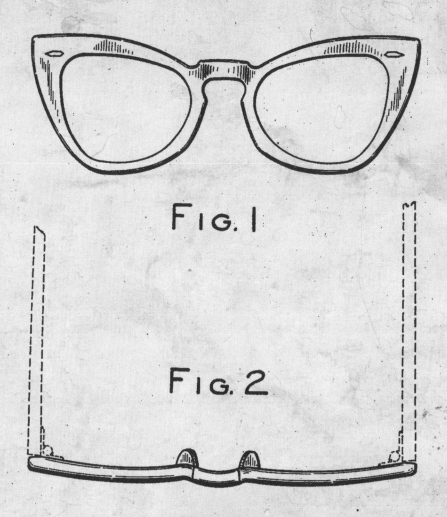

FIG. 1

FIG. 2

INVENTOR.
R.F.E. STEGEMAN
BY
S.A. Ellestad

ATTORNEY

1952

RAY-BAN WAYFARER

R. F. E. STEGEMAN

Responding to the demand for an update to the Ray-Ban range, hitherto essentially made of metal, the Wayfarer design represented a stylistic departure.

The sales of these sunglasses skyrocketed in 1961 after the release of the Blake Edwards film *Breakfast at Tiffany's*, in which they are worn by the superb Audrey Hepburn. Bob Dylan and Ray Charles contributed equally to their popularity.

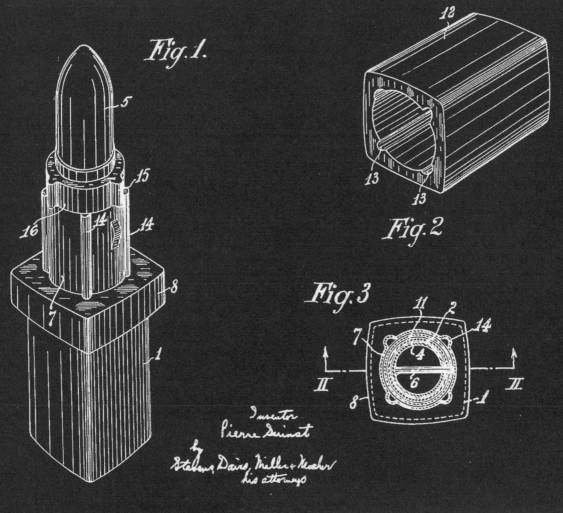

Fig. 1.

Fig. 2.

Fig. 3.

Inventor
Pierre Suinat
by
Stevens Davis, Miller + Mosher
his attorneys

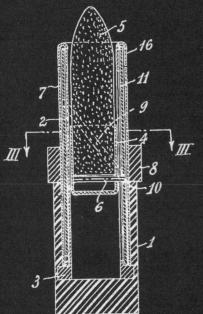

1952

LIPSTICK CASE

P. SUINAT

This item was invented and patented by Maurice Lévy, an American mechanical engineer, in 1915. The sliding/turning metal case serves as a sheath for the lipstick, which is thus protected from dust and does not mark the bag in which it is kept.

It was very quickly adopted and adapted in countless ways by cosmetics companies.

In 2009, the company Guerlain marketed a luxury edition of its Kiss Kiss lipstick in an 18-carat gold case encrusted with 199 diamonds.

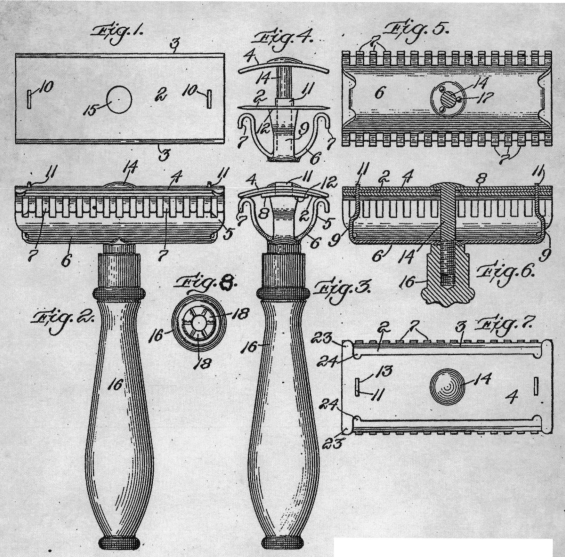

Fig. 1.

Fig. 4.

Fig. 5.

Fig. 2.

Fig. 8.

Fig. 3.

Fig. 6.

Fig. 7.

Witnesses:
Ruby M. Banfield.
Margaret H. Daniher.

Inventor:
King C. Gillette,

by

E. J. Chadwick,
Attorney.

1904

SAFETY RAZOR

K. C. GILLETTE

Made from 1903 onwards, this product was patented a year later. The object's major innovation is its system of interchangeable blades. The user must buy new blades regularly, which made the product a very lucrative one for Gillette.

During World War One, the brand provided 3.5 million razors and 36 million blades to the US Army.

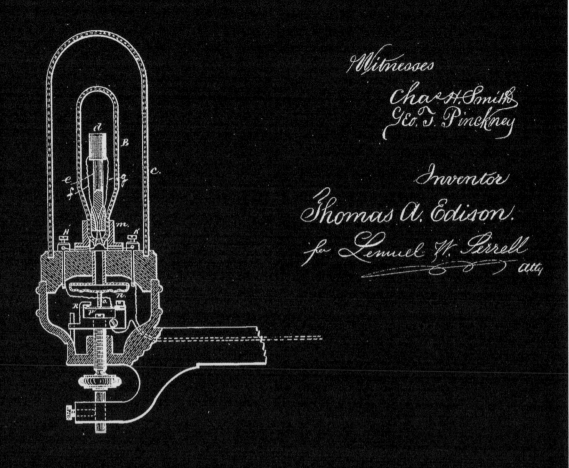

Witnesses
Chas. H. Smith
Geo. T. Pinckney

Inventor
Thomas A. Edison.
for Lemuel W. Serrell
atty.

1880

ELECTRIC LIGHT BULB

T. A. EDISON

It was while observing how the fibres of a piece of his fishing rod, thrown into the fire, glowed without disintegrating, that Thomas A. Edison had the idea of using organic matter to exploit the principle of incandescent illumination originally established by the Scotsman James Bowman Lindsay in 1835. Edison tested more than 6,000 samples of plant species gathered from around the world before finding that a carbonised bamboo filament was capable of lasting for more than 1,000 hours when an electric current was passed through it.

This patent, originally filed in April 1879, seals the filament within a vacuum inside the glass case, so as to preserve its longevity.

Fig. 1.

Fig. 2.

1888

GLOBE

W. M. GOLDTHWAITE

This globe benefits from an ingenious screw system which allows the axis of rotation to be changed at will, unlike earlier models which were limited to an angle of 23.5 degrees.

Although Greek astronomers created globes during antiquity, the oldest surviving terrestrial globe is the Erdapfel, or 'earth apple', created by the German cosmographer and navigator Martin Behaim between 1490 and 1492. Obviously, it doesn't feature either America or Australia.

Witnesses,

G. S. Robert
Geos Hughes Mead

Inventor:
William M. Goldthwaite,

By his Attorneys

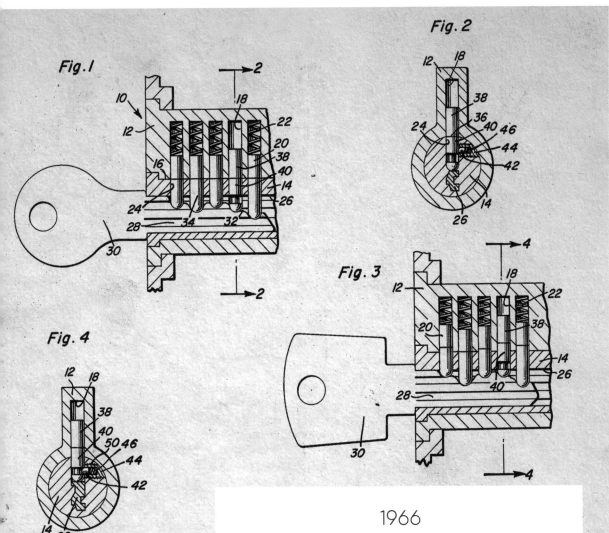

Fig. 1

Fig. 2

Fig. 3

Fig. 4

Fig. 5

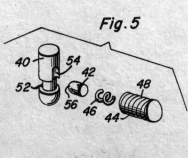

BY *Clarence A. O'Brien*
and Harvey B. Jacobson
                                  Attorneys

Ralph Bodek
Bennie A. DiFlavis
        INVENTORS

1966

PIN LOCK (PICK-PROOF LOCK)

R. BODEK

In 1861, the American Linus Yale Jr perfected an invention by his father for developing the cylindrical pin lock (also called the Yale lock), such as we know it today, although the principle goes back to ancient Egypt. In 1963, Ralph Bodek devised a version said to be 'pick-proof'.

Creating an inviolable lock is the dream of all locksmiths. In 1784, the British inventor Joseph Bramah devised a new locking system. To demonstrate its strength, he organised the 'Lock Challenge', offering 200 guineas to any craftsman who could succeed in making a tool capable of opening one of his padlocks.

This challenge remained undefeated for 67 years until an American locksmith by the name of Alfred C. Hobbs defeated the famous lock in 51 hours.

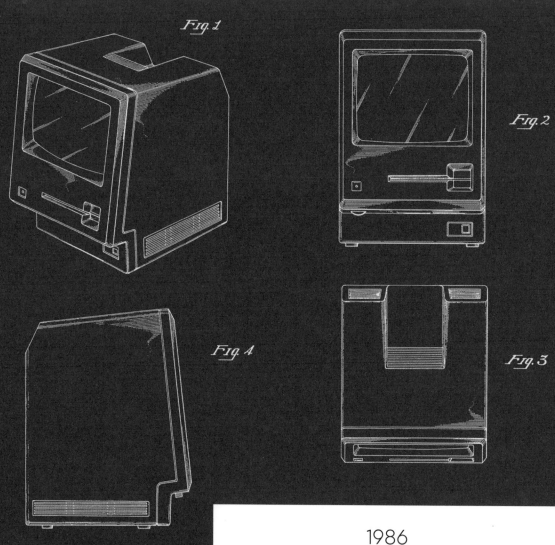

Fig. 1

Fig. 2

Fig. 4

Fig. 3

Fig. 6

1986

MACINTOSH SE

J. C. MANOCK

The launch of the first Macintosh computer, one of Apple's biggest successes, took place in 1984. These designs are for the SE (System Expansion) model, which appeared in 1987.

The first Apple logo, designed in 1976, showed Isaac Newton sitting under a tree, an apple hovering over his head. It was modernised a year later, taking the form of the apple with a bite taken out of it which we know today.

There are numerous hypotheses about this logo. Its designer, Rob Janoff, claims that the bite stops it being confused with a tomato, but it also plays on the word byte, the smallest unit of memory on a computer system.

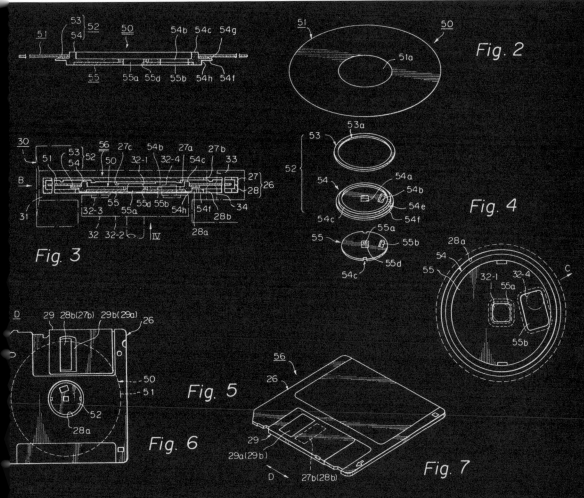

Fig. 2

Fig. 4

Fig. 3

Fig. 5

Fig. 6

Fig. 7

1992

FLOPPY DISK

FUJITSU LIMITED

The first floppy disks were created by IBM in 1967. They could hold up to 80,000 characters, the equivalent of a day's typing by a professional typist.

At the end of the twentieth century, their memory hadn't grown very much (up to 1.4 MB). However, they were gradually replaced by USB sticks, with a much greater capacity at a smaller physical size.

It wasn't until 2011 that Sony put a definitive stop to the production of 3.5-inch disks.

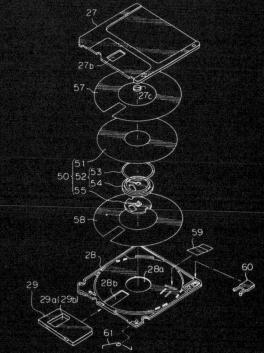

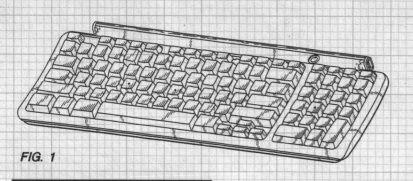

FIG. 1

FIG. 4

2000

APPLE KEYBOARD

S. JOBS

Invented to be sold with the iMac G3 in 1998, this keyboard did not have its patent validated and published until 2000.

The Apple keyboard has the special feature of integrating two USB 1.1 ports on either side. The arrangement of the keys most commonly used is the QWERTY, devised according to the frequency of use of the letters in English and originally created for the Sholes and Glidden typewriter. Other nations have used slight variations, such as the AZERTY in France and the QWERTZ in Germany.

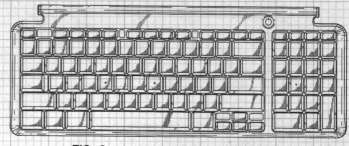

FIG. 2

FIG. 3

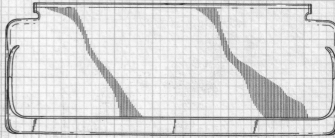

FIG. 5

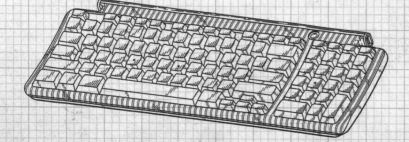

FIG. 6

FIG. 7

Fig. 1
Fig. 6
Fig. 3
Fig. 2
Fig. 8
Fig. 5
Fig. 9
Fig. 12

1984

APPLE MOUSE

W. LAPSON ET AL

The first model of the Apple mouse, provided with the Macintosh 128K, was designed in 1963 by Douglas Engelbart, a pioneering American computer engineer. It was nothing more at that stage than a wooden box equipped with two metal rollers and a button.

Its inventor only presented it five years later at a conference known as 'The Mother of All Demos'. A few years later, he gave up the licence of the Apple mouse for around $40,000.

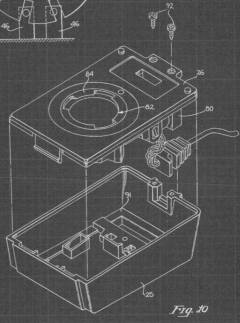

Fig. 10

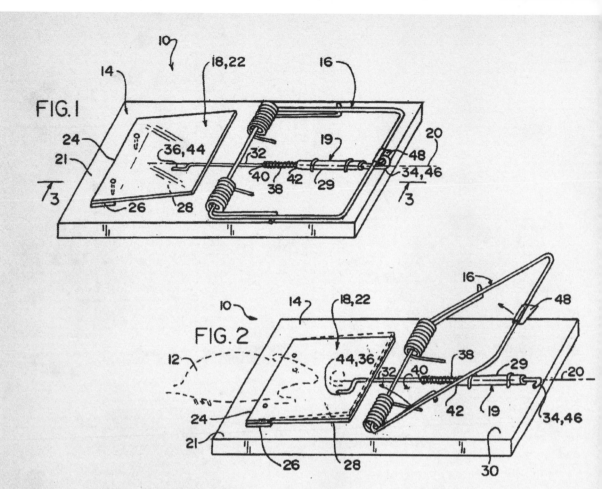

FIG.1

FIG.2

2003

MOUSETRAP

J. TREVINO

It was during the World's Fair of 1878 that Étienne Aurouze revealed his invention of the mousetrap, christened the 'spring trap'. He received numerous prizes for this innovation, including the World's Fair gold medal in 1894. The same year he opened his shop: The White Fox. This still belongs to the family of its founder and is today the last exterminator in the centre of Paris.

The patent of Jose Trevino is one of many descendants of the 1878 invention.

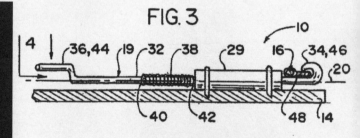

FIG.3

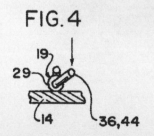

FIG.4

Fig. 1.

WITNESSES:

INVENTOR

George Kern

BY

Andrew Neureuther

ATTORNEY

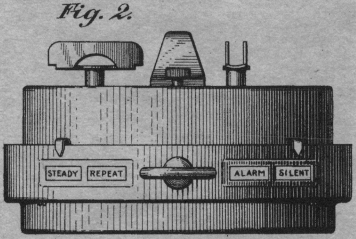

Fig. 2.

1911

ALARM CLOCK

G. KERN

The first alarms were invented several centuries ago. Plato developed a system based on a water clock to stop himself from nodding off when reading at night. In China in the tenth century, the fire clock was conceived, whereby a wick burns down and sets off a waking mechanism.

But it was in 1787 that Levi Hutchins, an apprentice clockmaker based in New Hampshire, USA, invented the alarm clock. Meanwhile, the first patent for an adjustable mechanical alarm was registered in 1847 by the Frenchman Antoine Redier.

George Kern, an inventor from Illinois, worked on numerous mechanisms and cases for the alarm clock but also on coin-operated machines.

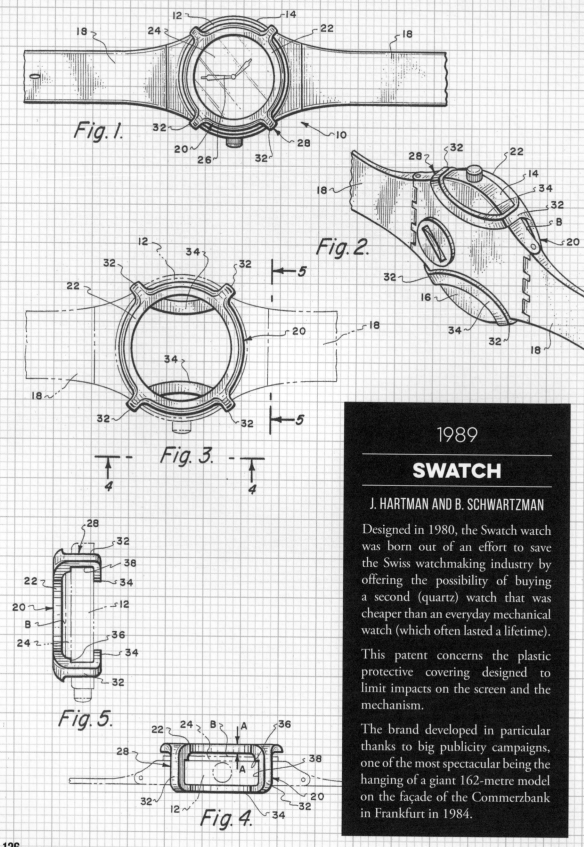

Fig. 1.

Fig. 2.

Fig. 3.

Fig. 5.

Fig. 4.

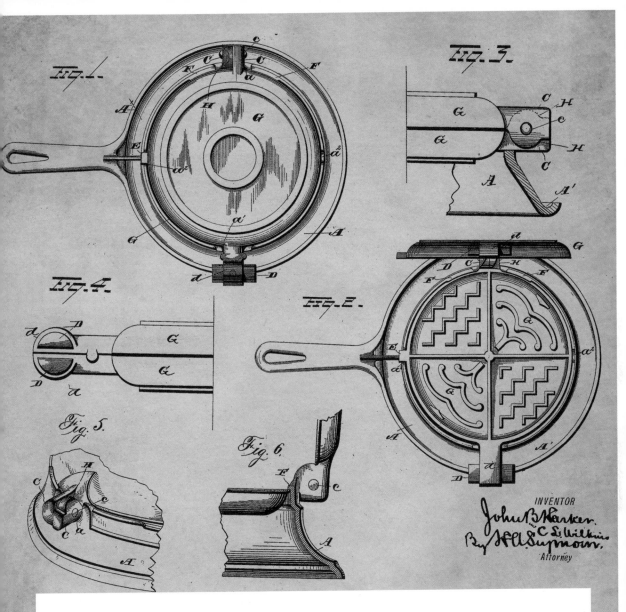

1883

WAFFLE IRON

J. B. HARKER AND C. L. WILKINS

We owe this invention to blacksmiths who imagined waffle irons inspired by wax honeycombs in beehives.

The type shown here is a classic nineteenth-century model, which chefs placed directly on the heat source. The decoration could vary and allowed for the personalisation of waffles with the shapes of flowers, fruits, birds, the sun, the moon, and so on.

Francis I, king of France in the early sixteenth century, was a great lover of waffles and had a waffle iron made in silver. It was struck with his initials and a salamander, his personal emblem.

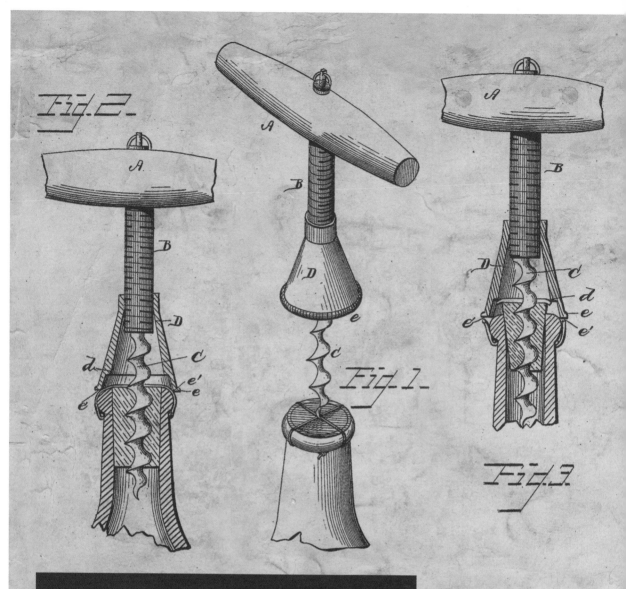

Fig.2

Fig.1

Fig.3

WITNESSES
F. L. Ourand
J. R. Littell

T. M. Strait
INVENTOR

by C. A. Snow & Co.
Attorneys

1883

CORKSCREW

T. M. STRAIT

The first types of corkscrew were developed in the eighteenth century in Britain. The need arose when the commercialisation of bottled wine began, and with it the use of cork stoppers.

The corkscrew was inspired by a device which installed a tap in a barrel to allow for the wine to be tasted during fermentation.

The first patent was granted rather late, in 1795, to the British clergyman Samuel Henshall of Oxford. Strait's version, shown here, had a shaft that moved down through a collar for ease of use.

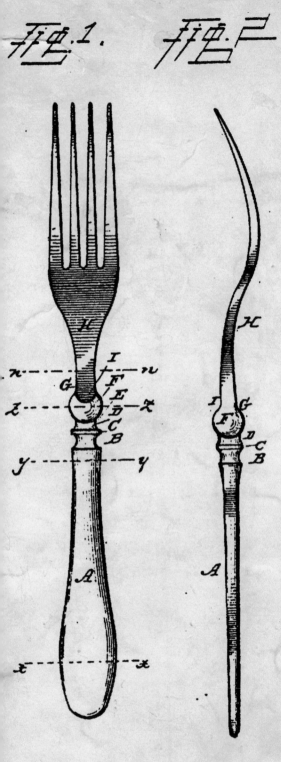

Fig. 1. Fig. 2. Fig. 3. Fig. 4. Fig. 5. Fig. 6.

1884

FORK

H. C. HART

In the eleventh century, a gold table fork was brought to Venice in the dowry of a Byzantine princess. It was considered a feminine accessory for centuries. In 1608, Thomas Coryate brought back from Italy the custom of eating meat with a fork, and in 1633 Charles I finally declared, 'It is decent to use a fork.'

This patent concerns a fork design made in Connecticut at the end of the nineteenth century.

WITNESSES:

Hubert. C. Hart
INVENTOR.

By

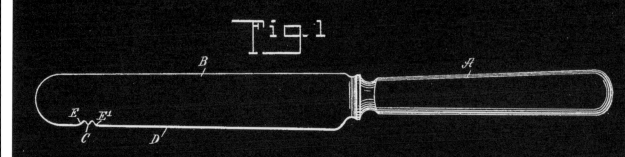

Fig.1

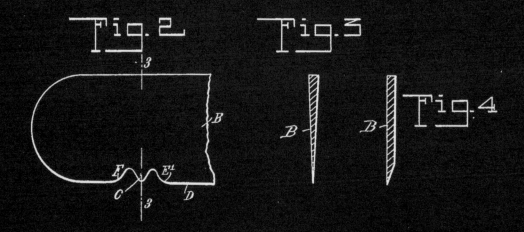

Fig.2   Fig.3

Fig.4

1911

TABLE KNIFE

H. E. CHANDLER

This knife with a rounded end, known as a table knife, shows one peculiarity: a 'tooth' supposed to facilitate the cutting of meat and fish.

According to tradition, in the seventeenth century Cardinal Richelieu was fed up of bad manners among his dinner guests, who would pick their teeth with the sharp end of their knives. He therefore instructed his steward to blunt the end, thus creating a table knife.

This innovation was encouraged by Louis XIV, who saw it as a way of eliminating dangerously sharp blades, a clear threat to his personal safety.

WITNESSES:
J. A. Brophy
Theof. Hoster

INVENTOR
Henry E. Chandler
BY Munroe
ATTORNEYS

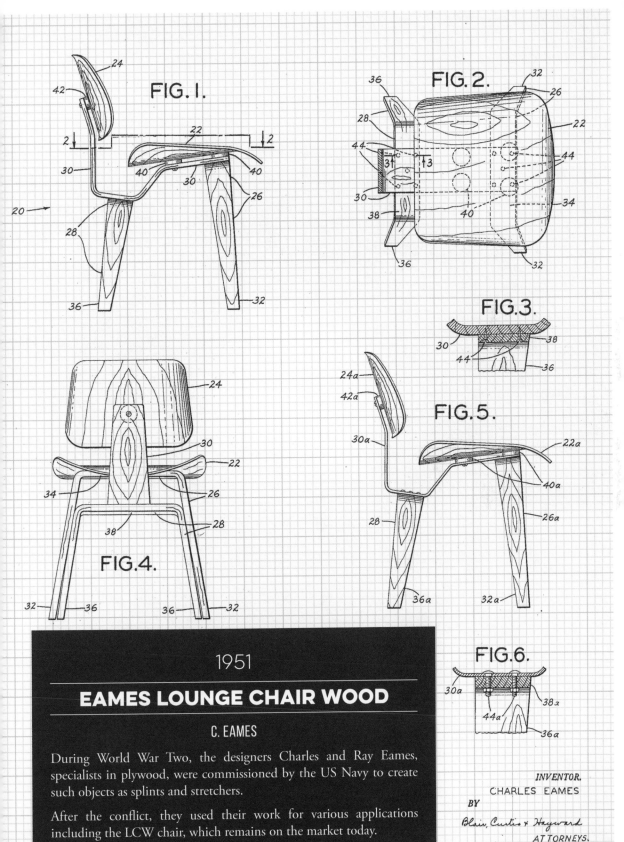

FIG. 1.

FIG. 2.

FIG.3.

FIG.5.

FIG.4.

FIG.6.

1951

EAMES LOUNGE CHAIR WOOD

C. EAMES

During World War Two, the designers Charles and Ray Eames, specialists in plywood, were commissioned by the US Navy to create such objects as splints and stretchers.

After the conflict, they used their work for various applications including the LCW chair, which remains on the market today.

INVENTOR.
CHARLES EAMES
BY
Blair, Curtis & Hayward
ATTORNEYS.

INDEX

If you're interested in finding out
more about our books, find us on Facebook
at **Summersdale Publishers** and follow us
on Twitter at @Summersdale.

www.summersdale.com